NEW ADVANCED HIGHER MATHEMATICS REVISION NOTES

Alexander Forrest

New Advanced Higher Mathematics
Revision Notes

Copyright © Alexander Forrest, 2019. All Rights Reserved.

All rights reserved. No part of this book may be reproduced in any form or by any electronic or mechanical means including information storage and retrieval systems, without permission in writing from the author. The only exception is by a reviewer, who may quote short excerpts in a review.

The right of Alexander Forrest to be identified as the author of this work has been asserted by him in accordance with the Copyright, Designs and Patents Act 1988.

Cover designed by Cover Designer

Printed by CreateSpace Independent Publishing Platform

First Printing: March 2019

Published in the United Kingdom by:
Beagle Bytes
beaglebytes@btinternet.com
Via the CreateSpace Independent Publishing Platform

Visit my website at www.mathsmutt.co.uk

Although every precaution has been taken in the preparation of this book, the publisher and author assume no responsibility for errors or omissions. Neither is any liability assumed for damages resulting from the use of this information contained herein

ISBN-13: 978-109181-753-1

For P

Contents

New Advanced Higher Mathematics 1
REVISION NOTES 1
    Methods in Algebra and Calculus 9
    Geometry, Proof and Systems of Equations 10
    Applications of Algebra and Calculus 11
Factorials and Binomial Theory 12
    Factorials 13
    Permutations and Combinations 15
    Properties of combinations 16
    Pascal's Triangle 17
    Binomial Theorem 19

Binomial expansions 19
    Probability and the binomial theorem 22
    The binomial expansion and e 23

Complex Numbers 24
    Sets reminder 25
    Complex Conjugate 26
    Arithmetic operations 27
    Argand diagrams 29
    Loci 32
    Polar form 33
    De Moivre's theorem 34
    Roots of a complex number 35
    Complex roots 37

Sequences and Series 38
    Sequences and Series 39
    Recurrence relations 39
    Fixed Points 40
    Arithmetic sequences 41
    Sum to n terms of an arithmetic sequence 43
    Geometric sequences 47
    Sum to n terms of a geometric sequence 49
    Infinite Series 51

- Arithmetic series 51
- Geometric series 52
- Sigma notation – Rules 54
- Sum of first n natural numbers 55
- Common SERIES – Sigma notation 57
- Power Series 58
- D'Alembert's ratio test 58
- Absolute convergence 60
- Fibonacci series 62
- Centre of convergence 62
- Taylor's series 65
- Maclaurin's series 65
- Standard series 72
- Iteration 74
- Newton Raphson Iteration 74
- Graphical methods 75
- First order process 81
- Using Taylor's series 81
- Second order process 82
- Rule of false position 83

Functions 85
- Modulus Function 86
- Odd and Even functions 87
- Asymptotes 90
- Finding asymptotes 91
- Critical Points 93

Matrices 97
- Types 98
- Elements 99
- Matrix notation 100
- Matrix Addition & Subtraction 101
- Scalar Matrix Multiplication 102
- Matrix Multiplication 102
- Transpose 105
- Special matrices 106
- Transformation Matrices 107
- Reflection and Rotation 108
- Enlargement with scale factor 109
- Determinants 110
- Determinant of a 3x3 matrix 111
- Cofactors 113
- Adjoint 116
- Inverse of a square matrix 117
- Product of a square matrix and its inverse 118
- Gaussian ELIMINATION 121

Vectors 124
- Direction Ratios and Cosines 125
- Vector Product 127
- Right Hand Screw rule. 127

Other properties 128
Components 129
Scalar Triple Product 133
Planes 135
Planes in space 140
The angle between two planes 141
The distance between parallel planes 142
Coplanar vectors 143
Vector equation of a plane 143
The equations of a line 145
The angle between a line and a plane 150
The intersection of two lines 152
The intersection of two planes 155
The distance from a point to a plane 157
The distance from a point to a line 159
The intersection of three planes 161

Partial Fractions 164

Partial Fractions 165
Distinct linear factors 166
Repeated linear factor 167
Irreducible quadratic factor 168

Calculus 169

Differentiation 170
Finding Gradients of Curves 171
Differentiation by First Principles   – deriving f'(x) from f(x) 178
Finding other derivatives by first principles 181
The Chain Rule 187
Product Rule : First Principles 188
Quotient Rule 189
Finding trigonometric derivatives by first principles 190
Gradients of tangents to curves 196
Increasing / Decreasing functions 197
Finding Stationary Points: Revision 199
Nature Tables: Revision 200
Second Derivative 201
Closed Intervals 202
Higher Derivatives 203
Differentiation Calculations Refresher 204
The Chain rule 207
Product Rule 208
Quotient Rule 209
Trig functions 210
Logarithms and Exponentials 212
Approximating roots of an equation : Newton's Method 213
Applications of differentiation 214
Rectilinear motion (straight line motion) 216
Differentiating Inverse functions 219
Differentiating Inverse Trig functions 223
Differentiating Explicit and Implicit Functions 227
Logarithmic Differentiation 230

- Parametric Equations 232
- Constraint equations 234
- Differentiating parametric equations 236
- Second derivative (PARAMETRICS) 238
- Other Uses 242

Integration 243

- Trig functions 245
- Logarithms and Exponentials 246
- Integration by substitution 247
- Common Forms 250
- Infinite Integrals 255
- The integrand becomes Infinite:- 257
- Infinite point within the integrand. 259
- Areas under curves 261
- Integration and the Area Function 266
- Definite integrals 267
- Area between curve and the y-axis 269
- Fundamental theorem of calculus 270
- Areas enclosed by the graph and the x – axis. 271
- Area between two graphs 272
- Formula for Area between two graphs 273
- Integrating inverse trig functions. 276
- Integration by Parts 277
- Integrating rational functions 279

Differential Equations 283

- First order Differential Equations 285
- Variables Seperable 287
- Applications of Differential Equations 290
- Rocket Science 290
- Newton's Law of Cooling 291
- Growth & Decay 292
- Kinematics 293
- Volumes of Solid of Revolution 294
- Applications of integration 297
- Linear Differential Equations 298
- First order Linear Differential Equations 299
- Second order Linear Differential Equations 302
- Second order non homogeneous Differential Equations 306

Number Theory 310

- The Division Algorithm 311
- The Euclidean Algorithm 312
- Diophantine Equations 317
- Pythagorean Triples 320
- Fermat's Theorem 320
- Number Bases 321
- Proof 323
- Proof has its own notation:- 326
- Direct Proof 327
- Indirect Proof (aka Proof by contradiction) 328

- Proof by Contrapositive 329
- Proof by Induction 330

Useful Formulae 337
- Set notation 338
- Common Sets 338
- Derivatives 339
- Standard Integrals 342
- **Trig formulae** 345
- Compound Formulae 347
- Double Angle Formulae   alf Angle Formulae 348
- Addition formulae   Products 348
- Multiple Angle Formulae 349
- Books by the Author 350

Index 351

METHODS IN ALGEBRA AND CALCULUS

Key subskills:

MAC 1.1 Applying algebraic skills to partial fractions.
- Expressing proper rational functions as a sum of partial fractions.

MAC 1.2 Applying calculus skills through techniques of differentiation.
- Differentiating functions given in the form of a product and in the form of a quotient.
- Differentiating exponential and logarithmic functions.
- Differentiating inverse trigonometric functions.
- Finding the derivative of functions defined implicitly.
- Finding the derivative of functions defined parametrically.

MAC 1.3 Applying calculus skills through techniques of integration.
- Integrating expressions using standard results.
- Integrating by substitution.
- Integrating by parts.

MAC 1.4 Applying calculus skills to solving differential equations.
- Solving a first order differential equation with variables separable.
- Solving a first order linear differential equation using the integrating factor.
- Solving second order differential equations.

GEOMETRY, PROOF AND SYSTEMS OF EQUATIONS

Key subskills:

GPS 1.1 Applying algebraic skills to matrices and systems of equations.
- Using Gaussian elimination to solve a 3x3 system of linear equations.
- Performing matrix operations of addition, subtraction and multiplication.
- Calculating the determinant of a matrix.
- Finding the inverse of a matrix.

GPS 1.2 Applying algebraic and geometric skills to vectors.
- Calculating a vector product.
- Finding the equation of a line in three dimensions.
- Finding the equation of a plane.

GPS 1.3 Applying geometric skills to complex numbers.
- Performing geometric operations on complex numbers

GPS 1.4 Applying algebraic skills to number theory.
- Using Euclid's algorithm to find the greatest common divisor of two positive integers.

GPS 1.5 Applying algebraic and geometric skills to methods of proof.
- Disproving a conjecture by providing a counter-example.
- Using direct and indirect proof in straightforward examples.

APPLICATIONS OF ALGEBRA AND CALCULUS

Key subskills:

Apps 1.1 Applying algebraic skills to the binomial theorem and to complex numbers
- Expand expressions using the binomial theorem.
- Performing algebraic operations on complex numbers.

Apps 1.2 Applying algebraic skills to sequences and series.
- Finding the general term and summing arithmetic and geometric sequences.
- Using the Maclaurin series expansion to find a stated number of terms of the power series for a simple function.

Apps 1.3 Applying algebraic skills to summation and mathematical proof.
- Applying summation formulae.
- Using proof by induction.

Apps 1.4 Applying algebraic and calculus skills to properties of functions.
- Finding the asymptotes of rational functions.
- Investigating features of graphs and sketching graphs of functions, including appropriate analysis of stationary points.

Apps 1.5 Applying algebraic and calculus skills to problems.
- Applying differentiation to problems, in context where appropriate.
- Applying integration to problems, in context where appropriate.

FACTORIALS AND BINOMIAL THEORY

Apps 1.1 : Applying algebraic skills to the binomial theorem and to complex numbers.

FACTORIALS

Example

Fred knows that the code to his locker is made up of the digits 5, 6, 7 and 8, but he can't remember the order. How many different possible codes exist to open his locker?

There are 4 ways of listing the first number,

5***   6***   7***   8***

and 3 ways of listing the second number.

56**	65**	75**	85**
57**	67**	76**	86**
58**	68**	78**	87**

This leaves 2 ways of selecting the third number

567*	657*	756*	856*
568*	658*	758*	857*
576*	675*	765*	865*
578*	678*	768*	867*
586*	685*	785*	875*
587*	687*	786*	876*

Finally, this leaves only one way of selecting the last number.

5678	6578	7568	8567
5687	6587	7586	8576
5768	6758	7658	8657
5786	6785	7685	8675
5867	6857	7856	8756
5876	6875	7865	8765

$0! = 1$ By definition
$1! = 1$
$2! = 2 \times 1 = 2$
$3! = 3 \times 2 \times 1 = 6$
$4! = 4 \times 3 \times 2 \times 1 = 24$
$5! = 5 \times 4 \times 3 \times 2 \times 1 = 120$

There are 24 possible codes.

$4 \times 3 \times 2 \times 1 = 24$

This is called 4-factorial and is written 4!

$n \times (n-1)! = n!$
so $(n-r+1) \times (n-r)! = (n-r+1)!$

Note, $5 \times 4! = 5 \times 24 = 120 = 5!$

In general, $$n! = n \times (n-1) \times (n-2) \times (n-3) \ldots \ldots \times 3 \times 2 \times 1$$

PERMUTATIONS AND COMBINATIONS

A permutation is an ordered arrangement of objects, in which order matters.
A combination is an arrangement of objects where order does not matter.

A group of n different objects has n! possible permutations.

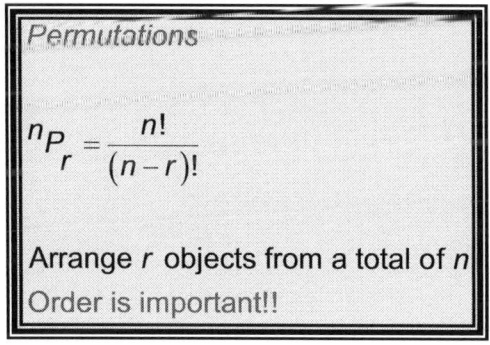

Permutations

$$^nP_r = \frac{n!}{(n-r)!}$$

Arrange r objects from a total of n
Order is important!!

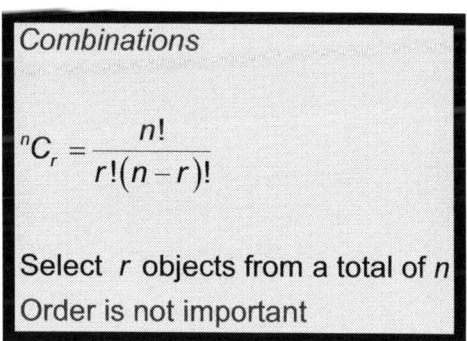

Combinations

$$^nC_r = \frac{n!}{r!(n-r)!}$$

Select r objects from a total of n
Order is not important

Example
There are 56 ways of choosing 5 different tins of cat food from 8 brands, but 6720 ways of ordering them!

$$^8C_5 = \frac{8!}{5!(8-5)!}$$
$$= \frac{8 \times 7 \times 6 \times 5 \times 4 \times 3 \times 2 \times 1}{5!3!}$$
$$= \frac{8 \times 7 \times 6 \times 5 \times 4 \times 3 \times 2 \times 1}{(5 \times 4 \times 3 \times 2 \times 1)(3 \times 2 \times 1)}$$
$$= \frac{8 \times 7 \times 6}{(3 \times 2 \times 1)}$$
$$= \frac{336}{6}$$
$$= 56$$

$$^8P_5 = \frac{8!}{(8-5)!}$$
$$= \frac{8 \times 7 \times 6 \times 5 \times 4 \times 3 \times 2 \times 1}{3!}$$
$$= \frac{8 \times 7 \times 6 \times 5 \times 4 \times 3 \times 2 \times 1}{3 \times 2 \times 1}$$
$$= 8 \times 7 \times 6 \times 5 \times 4$$
$$= 6720$$

PROPERTIES OF COMBINATIONS

$$^nC_n = \frac{n!}{n!(n-n)!}$$
$$= \frac{n!}{n!0!}$$
$$= \frac{n!}{n!}$$
$$= 1$$

$$^nC_0 = \frac{n!}{0!(n-0)!}$$
$$= \frac{n!}{0!n!}$$
$$= \frac{n!}{n!}$$
$$= 1$$

$$^nC_{n-r} = \frac{n!}{(n-r)!(n-(n-r))!}$$
$$= \frac{n!}{(n-r)!(n-n+r)!}$$
$$= \frac{n!}{(n-r)!(r)!}$$
$$= \frac{n!}{r!(n-r)!}$$
$$= {^nC_r}$$

PASCAL'S TRIANGLE

```
                              1
                           1  2  1
                         1  3   3  1
                       1  4   6   4  1
                     1  5  10  10   5  1
                   1  6  15  20  15   6  1
                 1  7  21  35  35  21   7  1
               1  8  28  56  70  56  28   8  1
             1  9  36  84 126 126  84  36   9  1
           1 10  45 120 210 252 210 120  45  10  1
```

The triangle is built by adding the row above.

	Column					
	0	1	2	3	4	5
row 0	1					
row 1	1	1				
row 2	1	2	1			
row 3	1	3	3	1		
row 4	1	4	6	4	1	
row 5	1	5	10	10	5	1

	Column					
	0	1	2	3	4	5
row 0	0C0					
row 1	1C0	1C1				
row 2	2C0	2C1	2C2			
row 3	3C0	3C1	3C2	3C3		
row 4	4C0	4C1	4C2	4C3	4C4	
row 5	5C0	5C1	5C2	5C3	5C4	5C5

This can be written ^nC_r, where n is the row, r is the column
Notice how $^3C_3 = {}^3C_0 = 1$ and $^3C_2 = {}^3C_1 = 3$

The coefficients are known as binomial coefficients.

Also, notice

$$^nC_r + {}^nC_{r+1} = {}^{n+1}C_{r+1} \qquad\qquad {}^nC_{r-1} + {}^nC_r = {}^{n+1}C_r$$

eg. $\ ^4C_2 + {}^4C_3 = 10 = {}^5C_3 \qquad\qquad$ eg. $\ ^4C_1 + {}^4C_2 = 10 = {}^5C_2$

^nC_r can be written in the form $\binom{n}{r}$

$$\binom{n}{r} + \binom{n}{r+1} = \binom{n+1}{r+1} \qquad\qquad \binom{n}{n-r} = \binom{n}{r}$$

$$\binom{n}{r-1} + \binom{n}{r} = \binom{n+1}{r} \qquad\qquad \binom{n}{n} = \binom{n}{0} = 1$$

Example

Solve the equation:

$$\binom{n}{n-2} = 55$$

$$\therefore \binom{n}{2} = 55$$

Look for 55 in column 2
of Pascal's triangle
This occurs in row 11.
n = 11

BINOMIAL THEOREM

Binomial expansions

Example

$$(x+y)^5$$
$$= 1x^5y^0 + 5x^4y^1 + 10x^3y^2 + 10x^2y^3 + 5x^1y^4 + 1x^0y^5$$
$$= {}^5C_0 x^5 y^0 + {}^5C_1 x^4 y^1 + {}^5C_2 x^3 y^2 + {}^5C_3 x^2 y^3 + {}^5C_4 x^1 y^4 + {}^5C_5 x^0 y^5$$
$$= \binom{5}{0}x^5y^0 + \binom{5}{1}x^4y^1 + \binom{5}{2}x^3y^2 + \binom{5}{3}x^2y^3 + \binom{5}{4}x^1y^4 + \binom{5}{5}x^0y^5$$

In General

$$(a+b)^n$$
$$= {}^nC_0 a^n b^0 + {}^nC_1 a^{n-1} b^1 + {}^nC_2 a^{n-2} b^2 + {}^nC_3 a^{n-3} by^3 + .. + {}^nC_r a^{n-r} b^r + .. + {}^nC_n a^{n-n} b^n$$
$$= \sum_{r=0}^{n} {}^nC_r a^{n-r} b^r$$

Alternatively, using $\binom{n}{r}$ instead of ${}^nC_r$

$$(a+b)^n$$
$$= \binom{n}{0}a^n b^0 + \binom{n}{1}a^{n-1}b^1 + \binom{n}{2}a^{n-2}b^2 + ... + \binom{n}{r}a^{n-r}b^r + ... + \binom{n}{n}a^{n-n}b^n$$
$$= \binom{n}{0}a^n + \binom{n}{1}a^{n-1}b + \binom{n}{2}a^{n-2}b^2 + \binom{n}{3}a^{n-3}b^3 + .. + \binom{n}{r}a^{n-r}b^r + .. + \binom{n}{n}b^n$$
$$= \sum_{r=0}^{n} \binom{n}{r} a^{n-r} b^r$$

This is known as the binomial theorem, and gives the expansion of $(a+b)^n$ where a and b are real numbers and n is a natural number.

The binomial coefficients are found in the nth row of Pascal's triangle.

Examples

Expand the following:

$$(1+2x)^3 = \binom{3}{0}1^3(2x)^0 + \binom{3}{1}1^2(2x)^1 + \binom{3}{2}1^1(2x)^2 + \binom{3}{3}1^0(2x)^3$$

$$= 1 + 3 \times (2x) + 3 \times (2x)^2 + (2x)^3$$

$$= 1 + 6x + 3 \times 4x^2 + 8x^3$$

$$= 1 + 6x + 12x^2 + 8x^3$$

$$(x-\frac{1}{x})^5 = \binom{5}{0}x^5\left(-\frac{1}{x}\right)^0 + \binom{5}{1}x^4\left(-\frac{1}{x}\right)^1 + \binom{5}{2}x^3\left(-\frac{1}{x}\right)^2 + \binom{5}{3}x^2\left(-\frac{1}{x}\right)^3 + \binom{5}{4}x^1\left(-\frac{1}{x}\right)^4 + \binom{5}{5}x^0\left(-\frac{1}{x}\right)^5$$

$$= x^5\left(-\frac{1}{x}\right)^0 + 5x^4\left(-\frac{1}{x}\right)^1 + 10x^3\left(-\frac{1}{x}\right)^2 + 10x^2\left(-\frac{1}{x}\right)^3 + 5x^1\left(-\frac{1}{x}\right)^4 + \left(-\frac{1}{x}\right)^5$$

$$= x^5 - 5x^3 + 10x - \frac{10}{x} + \frac{5}{x^3} - \frac{1}{x^5}$$

The binomial theorem allows a specific term to be found from the general form.

$$(a+b)^n$$
$$= {}^nC_0 a^n b^0 + {}^nC_1 a^{n-1} b^1 + {}^nC_2 a^{n-2} b^2 + {}^nC_3 a^{n-3} by^3 + .. + {}^nC_r a^{n-r} b^r + .. + {}^nC_n a^{n-n} b^n$$
$$= a^n + na^{n-1}b + \frac{n(n-1)}{2!}a^{n-2}b^2 + \frac{n(n-1)(n-2)}{3!}a^{n-3}b^3 + ... + b^n$$

$(a+b)^n$ has general term $\quad {}^nC_r a^{n-r} b^r$

to find any specific term, set r.

Example, when r = 3

$${}^nC_3 a^{n-3} b^3 = \frac{n!}{3!(n-3)!} a^{n-3} b^3$$

$$= \frac{n(n-1)(n-2)}{3!} a^{n-3} b^3$$

Which is the fourth term of the expansion !

${}^nC_r a^{n-r} b^r$ is the $(r+1)$th term in the expansion for $0 \leq r \leq n$

Examples

Find the seventh term in the expansion of $(2x + 3y)^9$

$(2x + 3y)^9$ has general term $\ ^nC_r a^{n-r} b^r$

a = 2x , b = 3y, n=9

r+1=7, so r = 6

$$^9C_6 (2x)^{9-6} (3y)^6 = \frac{9!}{6!3!} \times (2x)^3 \times (3y)^6$$
$$= 84 \times 2^3 \times 3^6 \times x^3 \times y^6$$
$$= 489888 x^3 y^6$$

Find the term containing x^3 in the expansion of $(3 + 2x)^5$

$(3 + 2x)^5$ has general term $\ ^nC_r a^{n-r} b^r$

a = 3 , b = 2x, n=5

want term containing x^3, so $r = 3$

$$^5C_3 3^{5-3} (2x)^3 = \frac{5!}{3!2!} \times 3^2 \times 2^3 \times y^3$$
$$= 10 \times 9 \times 8 \times y^3$$
$$= 720 x^3$$

$$(a+b)^n = \sum_{r=0}^{n} {^nC_r} a^{n-r} b^r$$

when $a = 1$ and $b = 1$

$$(1+1)^n = \sum_{r=0}^{n} {^nC_r} 1^{n-r} 1^r$$

$$= \sum_{r=0}^{n} {^nC_r}$$

$$\therefore\ 2^n = \sum_{r=0}^{n} {^nC_r}$$

The sum of the numbers in Pascal's triangle is equal to 2^n, where n is the row number.

PROBABILITY AND THE BINOMIAL THEOREM

Let p denote the probability of success of an outcome
and q denote the probability of failure

Then $p + q = 1$

The probability of obtaining r successes in n independant trials is

$P(r \text{ successes}) = {}^nC_r q^{n-r} p^r$

Example

The probability that it will rain on any given day is 0.4.
What is the probability that it will rain twice in a seven day week?

$p = 0.4$
$q = 0.6$

$n = 7$ (no of possible outcomes)
$r = 2$ (no of required outcomes)

$$\begin{aligned}
P(\text{rain twice}) &= {}^nC_r q^{n-r} p^r \\
&= {}^7C_2\, 0.6^5\, 0.4^2 \\
&= \frac{7!}{2!5!} \times 0.6^5 \times 0.4^2 \\
&= 21 \times 0.07776 \times 0.16 \\
&= 0.26127 \\
&= 26.1\%
\end{aligned}$$

THE BINOMIAL EXPANSION AND E

$$\left(1+\frac{1}{n}\right)^n = 1 + n\left(\frac{1}{n}\right) + \frac{n(n-1)}{2!}\left(\frac{1}{n}\right)^2 + \ldots + \left(\frac{1}{n}\right)^n$$

$$\lim_{n \to \infty}\left(\frac{1}{n}\right) = 0$$

$$\left(1+\frac{1}{n}\right)^n = \sum_{r=0}^{\infty} \frac{1}{r!}$$

$$\sum_{r=0}^{\infty} \frac{1}{r!} = e$$

$$e^x = \sum_{r=0}^{\infty} \frac{x^r}{r!}$$

$$e^x = 1 + x + \frac{x^2}{2!} + \frac{x^3}{3!} + \ldots\ldots\ldots\ldots$$

Examples

$$e^3 = 1 + 3 + \frac{3^2}{2!} + \frac{3^3}{3!} + \frac{3^4}{4!} + \frac{3^5}{5!} + \frac{3^6}{6!} + \frac{3^7}{7!} + \ldots\ldots\ldots\ldots$$

$$e^3 = 1 + 3 + \frac{9}{2} + \frac{27}{6} + \frac{81}{24} + \frac{243}{120} + \frac{729}{720} + \frac{2187}{5040}\ldots\ldots\ldots$$

$$e^3 = 1 + 3 + 4.5 + 4.5 + 3.375 + 2.025 + 1.0125 + 0.4339\ldots\ldots$$

COMPLEX NUMBERS

GPS 1.3: Applying geometric skills to complex numbers.

Apps 1.1: Applying algebraic skills to the binomial theorem and to complex numbers.

SETS REMINDER

$\mathbb{N}$ is the set of natural numbers used in counting: $\{1,2,3,4.....\infty\}$

$\mathbb{W}$ is the set of whole numbers: $\{0,1,2,3,4.....\infty\}$

$\mathbb{Z}$ is the set of integers:- positive and negative whole numbers
$\{-\infty,...-4,-3,-2,-,1,0,1,2,3,4......\infty\}$

$\mathbb{Q}$ is the set of rational numbers or quotients.

These are all numbers which can be expressed as a fraction, $\dfrac{a}{b}$ where both a and b are integers, and b is not zero.

$\mathbb{R}$ is the set of real numbers $\{-\infty,........\infty\}$
This includes all numbers, rational and irrational.

$\mathbb{N} \subset \mathbb{W} \subset \mathbb{Z} \subset \mathbb{Q} \subset \mathbb{R}$

$\mathbb{C}$ is the set of complex numbers, $a + b\text{i}$
where a and b are real and i is the imaginary number $\sqrt{-1}$

COMPLEX CONJUGATE

Given $z = a + bi$

a is called the real part of z
$a = \Re(z)$ or $a = \text{Re}(z)$

b is called the imaginary part of z
$b = \vartheta(z)$ or $b = \text{Im}(z)$

$z = a + bi$ has complex conjugate $\bar{z} = a - bi$
$z\bar{z} = (a + bi)(a - bi) = a^2 + b^2$

Example

Given $z = 2 - 3i$
Find a) $\bar{z}$ b) $z\bar{z}$

a) $\bar{z} = 2 + 3i$
b) $z\bar{z} = (2 - 3i)(2 + 3i) = 4 + 9 = 13$

equating real and imaginary parts
If $z_1 = a + bi$ and $z_2 = c + di$
and $z_1 = z_2$,
$\Re(z_1) = \Re(z_2)$ and $\vartheta(z_1) = \vartheta(z_2)$

ARITHMETIC OPERATIONS

Given $z_1 = a+bi$ and $z_2 = c+di$

$z_1 + z_2 = (a+c)+(b+d)i$

$z_1 - z_2 = (a-c)+(b-d)i$

$z_1 z_2 = (a+bi)(c+di)$
$= ac+bci+adi+bdi^2$
$= ac+(bc+ad)i+bd\times(-1)$
$= (ac-bd)+(bc+ad)i$

$\dfrac{z_1}{z_2} = \dfrac{a+bi}{c+di} = \dfrac{a+bi}{c+di} \times \dfrac{c-di}{c-di}$

$= \dfrac{(a+bi)(c-di)}{(c+di)(c-di)}$

$= \dfrac{ac+bci-adi+bd}{c^2+d^2}$

$= \dfrac{(ac+bd)+(bc-ad)i}{c^2+d^2}$

Examples
Solve the equation $x^2 - 2x + 5 = 0$

$x^2 - 2x + 5 = 0$

$x = \dfrac{-b \pm \sqrt{b^2 - 4ac}}{2a}$

$= \dfrac{2 \pm \sqrt{4^2 - 20}}{2}$

$= \dfrac{2 \pm \sqrt{-4}}{2}$

$= \dfrac{2 \pm 2i}{2}$

$= 1 \pm i$

Given $z_1 = 2 - 3i$ and $z_2 = 1 + i$

Find

a) $z_1 + z_2$ b) $z_1 - z_2$ c) $z_1 z_2$ d) $\dfrac{z_1}{z_2}$

a) $z_1 + z_2 = (2 + 1) + (-3 + 1)i = 3 - 2i$

b) $z_1 - z_2 = (2 - 1) + (-3 - 1)i = -1 - 4i = -(1 + 4i)$

c) $z_1 z_2 = (2 - 3i)(1 + i)$
$= 2 - 3i + 2i + 3$
$= 5 - i$

d) $\dfrac{z_1}{z_2} = \dfrac{2 - 3i}{1 + i} = \dfrac{2 - 3i}{1 + i} \times \dfrac{1 - i}{1 - i}$

$= \dfrac{2 - 3i - 2i - 3}{1^2 + (-1)^2}$

$= \dfrac{-1 - 5i}{2}$

Alternatively

d)

$z_1 = 2 - 3i$ and $z_2 = 1 + i$

$a = 2 \qquad c = 1$
$b = -3 \qquad d = 1$

$\dfrac{z_1}{z_2} = \dfrac{(ac + bd) + (bc - ad)i}{c^2 + d^2}$

$= \dfrac{(2 - 3) + (-3 - 2)i}{1^2 + (-1)^2}$

$= \dfrac{-1 - 5i}{2}$

Inverse

Given $z = 2 - 3i$ Find z^{-1}

$z = 2 - 3i$

$\Rightarrow z^{-1} = \dfrac{1}{2 - 3i}$

$= \dfrac{1}{2 - 3i} \times \dfrac{2 + 3i}{2 + 3i}$

$= \dfrac{2 + 3i}{4 + 9}$

$= \dfrac{2 + 3i}{13}$

ARGAND DIAGRAMS

z = x +iy can be represented on the complex plane by the point P(x,y).

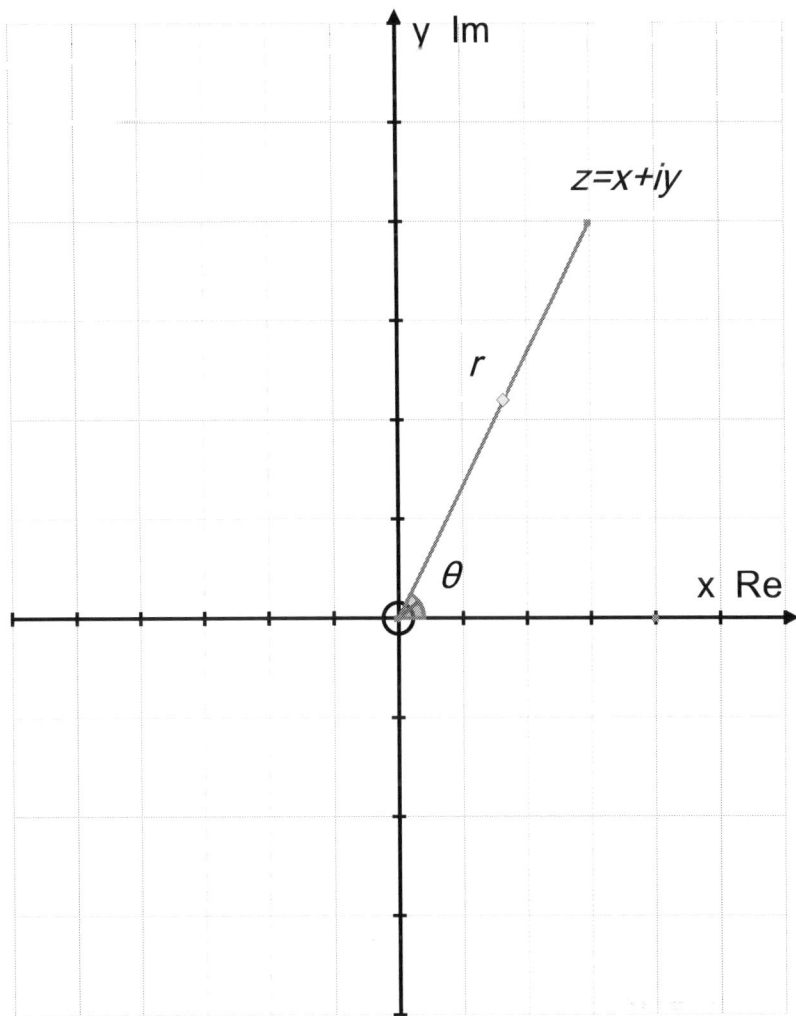

Points on the x axis are real.

Points on the y axis are imaginary.

$z = x + iy$ can also be represented by the vector $\overrightarrow{OP}$

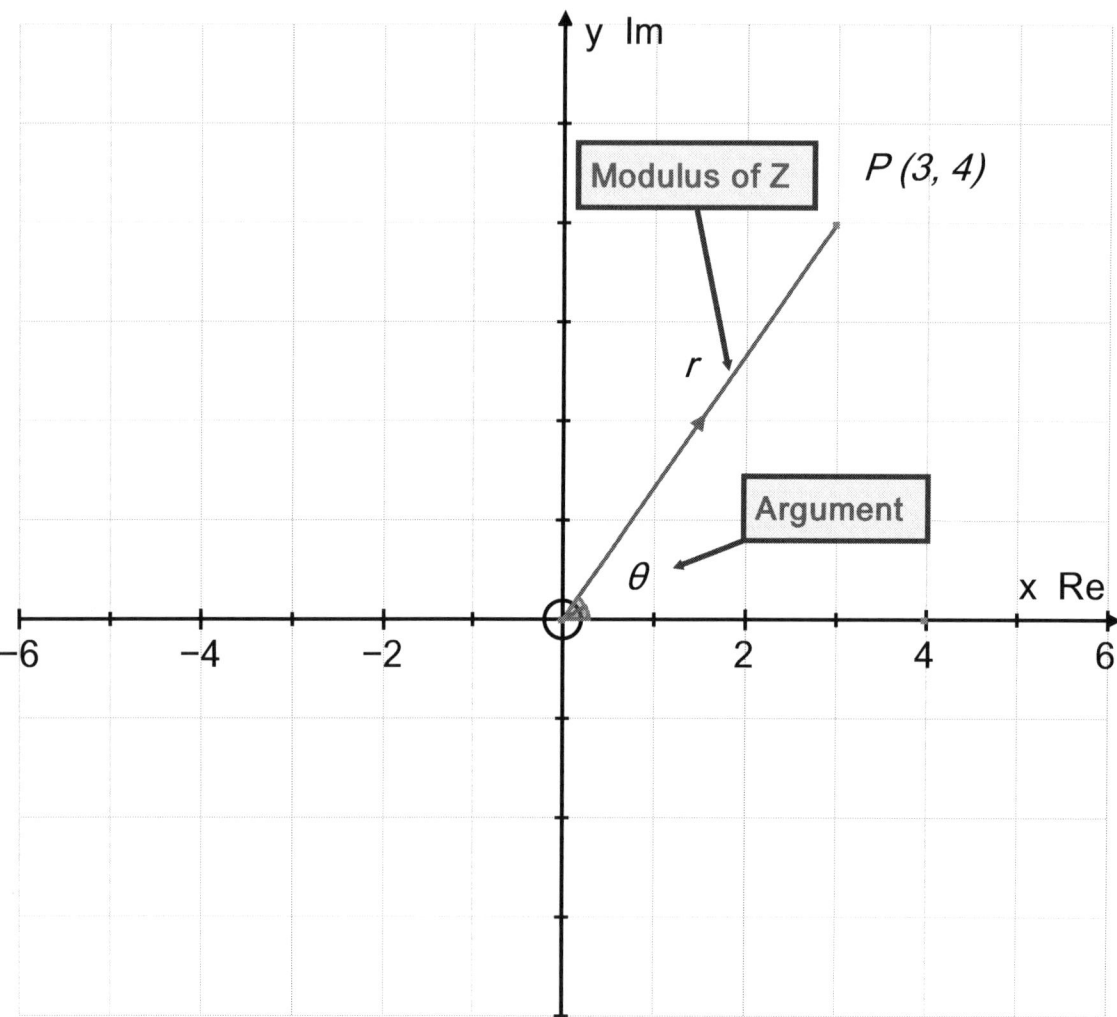

The length of $\overrightarrow{OP}$, r, is called the modulus of z.

$r = \overrightarrow{OP} = |z|$

By Pythagoras' Theorem

$$r = \sqrt{x^2 + y^2}$$
$$\therefore |z| = \sqrt{x^2 + y^2}$$

The size of the rotation is called the amplitude or argument of z.

Arg z = θ + 2nπ

$$\sin\theta = \frac{y}{r} \qquad \cos\theta = \frac{x}{r}$$
$$\Rightarrow y = r\sin\theta \qquad x = r\cos\theta$$

$$\tan\theta = \frac{\sin\theta}{\cos\theta} = \frac{y}{x}$$

$$\Rightarrow \theta = \tan^{-1}\left(\frac{y}{x}\right)$$

$$Arg\ z = \tan^{-1}\left(\frac{y}{x}\right) + 2n\pi \text{ radians}$$

The principal argument is denoted arg z and lies in the range $-\pi < \theta \leq \pi$

Example

Find the modulus and argument of the complex number z = 3 +4i

$$|z| = \sqrt{x^2 + y^2} \qquad Arg\ z = \tan^{-1}\left(\frac{y}{x}\right)$$
$$= \sqrt{3^2 + 4^2}$$
$$= \sqrt{25} \qquad\qquad = \tan^{-1}\left(\frac{4}{3}\right)$$
$$= 5$$
$$\qquad\qquad\qquad = 0.927 + n\pi \text{ rads}$$

Since the point is in the first quadrant, $n = 0$

arg z = 0.927 rads

LOCI

Given that z = x + iy, find the equation of the locus of :-

$|z| = 3$

$|z| = 3$
$\Rightarrow \sqrt{x^2 + y^2} = 3$
$\Rightarrow x^2 + y^2 = 9$

This describes a circle centre(0,0) radius 3

$|z - 5| = 8$

$|z - 5| = 8$
$\Rightarrow |x - 5 + iy| = 8$
$\Rightarrow \sqrt{(x-5)^2 + y^2} = 8$
$\Rightarrow (x-5)^2 + y^2 = 64$

This describes a circle centre (5,0) radius 8

$|z - 5i| = 8$

$|z - 5i| = 8$
$\Rightarrow |x + iy - 5i| = 8$
$\Rightarrow \sqrt{x^2 + (y-5)^2} = 8$
$\Rightarrow x^2 + (y-5)^2 = 64$

This describes a circle centre (0,5) radius 8

$|2z - 3i| = 5$

$|2z - 3i| = 5$
$\Rightarrow |2(x + iy) - 3i| = 5$
$\Rightarrow |2x + 2iy - 3i| = 5$
$\Rightarrow |2x + i(2y - 3)| = 5$
$\Rightarrow \sqrt{2x^2 + (2y-3)^2} = 5$
$\Rightarrow 2x^2 + (2y-3)^2 = 25$

POLAR FORM

$y = r\sin\theta \qquad x = r\cos\theta$

$z = x + iy$ becomes

$z = r\cos\theta + r\sin\theta i$

$ = r(\cos\theta + i\sin\theta)$

$$\boxed{z = r(\cos\theta + i\sin\theta)}$$

Example

Express z = 3 +4i in polar form.

$r = |z| = \sqrt{x^2 + y^2}$
$ = \sqrt{3^2 + 4^2}$
$ = \sqrt{25}$
$ = 5$

$\theta = \arg z = \tan^{-1}\left(\dfrac{y}{x}\right)$
$ = \tan^{-1}\left(\dfrac{4}{3}\right)$
$ = 0.927$

$z = r(\cos\theta + i\sin\theta)$
$ = 5(\cos 0.927 + i\sin 0.927)$

Note

$$\boxed{\begin{array}{lllll} |z|^2 = z\bar{z} & |\bar{z}| = |z| & |z_1 z_2| = |z_1| \times |z_2| & \left|\dfrac{z_1}{z_2}\right| = \dfrac{|z_1|}{|z_2|} \\[2ex] z^{-1} = \dfrac{\bar{z}}{|z|^2}, \; z \neq 0 & & \left|\dfrac{1}{z}\right| = \dfrac{1}{|z|} & \end{array}}$$

$$\boxed{\arg\left(\dfrac{1}{z}\right) = -\arg z \qquad \mathrm{Arg}\,(z_1 z_2) = \mathrm{Arg}\,z_1 + \mathrm{Arg}\,z_2 \qquad \mathrm{Arg}\left(\dfrac{z_1}{z_2}\right) = \mathrm{Arg}\,z_1 + \mathrm{Arg}\,z_2}$$

DE MOIVRE'S THEOREM

$$\boxed{z^n = r^n(\cos n\theta + i\sin n\theta)}$$

Example

Given z = 3 + 4i , calculate z^5

$$r = |z| = \sqrt{x^2 + y^2}$$
$$= \sqrt{3^2 + 4^2}$$
$$= \sqrt{25}$$
$$= 5$$

$$\theta = \arg z = \tan^{-1}\left(\frac{y}{x}\right)$$
$$= \tan^{-1}\left(\frac{4}{3}\right)$$
$$= 0.927$$

$$z = r(\cos\theta + i\sin\theta)$$
$$= 5(\cos 0.927 + i\sin 0.927)$$

$$z^5 = \left(5(\cos 0.927 + i\sin 0.927)\right)^5$$
$$= 5^5(\cos(5 \times 0.927) + i\sin(5 \times 0.927))$$
$$= 3125(\cos(4.635) + i\sin(4.635))$$
$$= 3125(-0.0773 - 0.997i)$$
$$= -241.5625 - 3115.625i$$
$$= -242 - 3116i \quad \text{(nearest integer)}$$

ROOTS OF A COMPLEX NUMBER

$$z = r(\cos\theta + i\sin\theta)$$
has n solutions $z_1^n = z$

$$z_1 = r^{\frac{1}{n}}\left(\cos\left(\frac{\theta+2k\pi}{n}\right) + i\sin\left(\frac{\theta+2k\pi}{n}\right)\right)$$
$$k = 0, 1, 2, 3 \ldots n-1$$

Example

Solve the equation $z^4 = -3 + 3\sqrt{3}i$

$$z^4 = -3 + 3\sqrt{3}i$$
$$|z^4| = \sqrt{(-3)^2 + (3\sqrt{3})^2}$$
$$= \sqrt{36}$$
$$= 6$$

$$\theta = \arg z = \tan^{-1}\left(\frac{y}{x}\right)$$
$$= \tan^{-1}\left(\frac{3\sqrt{3}}{-3}\right)$$
$$= \tan^{-1}\left(-\sqrt{3}\right)$$
$$= \frac{2\pi}{3}$$

$z^4 = -3 + 3\sqrt{3}i$

has solutions of the form

$$z = 6^{\frac{1}{4}}\left(\cos\left(\frac{1}{4}\left(\frac{2\pi}{3} + 2k\pi\right)\right) + i\sin\left(\frac{1}{4}\left(\frac{2\pi}{3} + 2k\pi\right)\right)\right), k = 0,1,2,3$$

when $k = 0$

$$z = 6^{\frac{1}{4}}\left(\cos\left(\frac{\pi}{6}\right) + i\sin\left(\frac{\pi}{6}\right)\right)$$

when $k = 1$

$$z = 6^{\frac{1}{4}}\left(\cos\left(\frac{2\pi}{3}\right) + i\sin\left(\frac{2\pi}{3}\right)\right)$$

when $k = 2$

$$z = 6^{\frac{1}{4}}\left(\cos\left(\frac{7\pi}{6}\right) + i\sin\left(\frac{7\pi}{6}\right)\right)$$

when $k = 3$

$$z = 6^{\frac{1}{4}}\left(\cos\left(\frac{19\pi}{6}\right) + i\sin\left(\frac{19\pi}{6}\right)\right)$$

COMPLEX ROOTS

If the root of a polynomial is unreal, it has complex roots.

r(cosθ +isinθ) and r(cosθ - isinθ)

A polynomial of degree n will have n complex roots.

Example

Find the roots of the equation

$z^3 - 6z^2 + 13z - 20 = 0$, given $z = 1+2i$ is a root.

If z=1+2i is a root, then so is its conjugate z = 1- 2i

Factors are z -1 -2i and z -1+2i

$(z-1-2i)(z-1+2i) = z^2 - 2z + 5$

$$\begin{array}{r} z-4 \\ z^2-2z+5 \overline{\smash{\big)}\, z^3-6z^2+13z-20} \\ \underline{z^3-2z^2+5z} \\ -4z^2+8z-20 \end{array}$$

The roots are $z = 4, z = 1 - 2i \text{ and } z = 1 + 2i$

SEQUENCES AND SERIES

Apps 1.2: Applying algebraic skills to sequences and series.

Apps 1.3: Applying algebraic skills to summation and mathematical proof.

SEQUENCES AND SERIES

Refresher

2, 3, 4, 5, 6, ..., n is the sequence created by the rule n + 1

u_n denotes the nth term of the sequence so here, $u_1 = 2$, $u_2 = 3$ etc.
A sequence can be formed by a recurrence relation.

A first-order linear recurrence relation is of the form $u_{n+1} = ru_n + d$, where r and d are constants.

A series is a sum formed by the terms of a sequence.

RECURRENCE RELATIONS

Example
Find the first-order linear recurrence relation given by the sequence 2, 4, 10, 28.
Here, $u_1=2$, $u_2=4$, $u_3=10$, $u_4=28$
$u_{n+1} = ru_n + d$

so, $4 = 2r + d$ and $10 = 4r + d$
therefore, $6 = 2r$, thus $r = 3$

Substituting back, $4 = 2r + d$
so $4 = 6 + d$
giving $d = -2$

This gives the first-order linear recurrence relation
$u_{n+1} = 3u_n - 2$

FIXED POINTS

If the previous relation is used with a starting value of $u_1=1$, then the following sequence is generated: -

1,1,1, 1,.......1 and so on.

Such repetition is called a fixed point.

$u_1=0$ generates the sequence

-2, -8, -26 etc.

Since $u_1=0$ and $u_1=2$ generate sequences which diverge from 1,

$u_n=1$ is an unstable fixed point of the relation $u_{n+1} = 3u_n - 2$

A stable fixed point occurs when the other sequences converge to the fixed point.

In general

$u_{n+1} = ru_n + d$ has a fixed point when $u_{n+1} = u_n$

$$\begin{aligned} u_{n+1} &= ru_n + d \\ \Rightarrow u_n &= ru_n + d \\ \Rightarrow u_n - ru_n &= d \\ \Rightarrow u_n(1-r) &= d \\ \Rightarrow u_n &= \frac{d}{(1-r)} \end{aligned}$$

For a stable fixed point $u_n = \dfrac{d}{(1-r)}$, $r \neq 1$

$|r| < 1$

A stable fixed point is often called a limit.

ARITHMETIC SEQUENCES

An arithmetic sequence is of the form

$u_{n+3} - u_{n+2} = u_{n+2} - u_{n+1} = u_{n+1} - u_n = d$

The common difference, d, is a constant.
The nth term of an arithmetic sequence can be described

$$u_n = a + (n-1)d$$

where a = u_1, d = common difference and n = n^{th} term.

Examples

Find the nth term of the arithmetic sequence   13, 16, 19...

a = 13 , d = 16 - 13 = 3

$u_n = a + (n-1)d$
$\Rightarrow u_n = 13 + (n-1) \times 3$
$\Rightarrow u_n = 13 + 3n - 3$
$\Rightarrow u_n = 10 + 3n$
$\Rightarrow u_n = 3n + 10$

Find the 219th term of the arithmetic sequence   13, 16, 19...
From above,

$u_n = 3n + 10$
$\Rightarrow u_{219} = 3 \times 219 + 10$
$\Rightarrow u_{219} = 657 + 10$
$\Rightarrow u_{219} = 667$

Find the arithmetic sequence which includes $u_7 = 33$ and $u_{18} = 88$

$u_n = a + (n-1)d$ and $u_n = a + (n-1)d$
$\Rightarrow 33 = a + (7-1)d$ $\Rightarrow 88 = a + (18-1)d$
$\Rightarrow 33 = a + 6d$ $\Rightarrow 88 = a + 17d$

so
$33 = a + 6d$
$88 = a + 17d$

solve simultaneously
$55 = 11d$
$\Rightarrow d = 5$

substitute
$33 = a + 6d$
$\Rightarrow 33 = a + 30$
$\Rightarrow a = 3$

$u_n = 3 + (n-1) \times 5$
$\Rightarrow u_n = 3 + 5(n-1)$
$\Rightarrow u_n = 3 + 5n - 5$
$\Rightarrow u_n = 5n - 2$

SUM TO N TERMS OF AN ARITHMETIC SEQUENCE

$$S_n = \frac{1}{2}n(2a+(n-1)d)$$

Examples

Find the sum of the first 20 terms of the arithmetic sequence that starts

12, 22, 32, 42, 52....

$a = 12$, $d = 10$, $n = 20$

$$S_n = \frac{1}{2}n(2a+(n-1)d)$$

$\Rightarrow S_{20} = \frac{1}{2} \times 20(2 \times 12 + (20-1) \times 10)$

$\Rightarrow S_{20} = \frac{1}{2} \times 20(24 + 190)$

$\Rightarrow S_{20} = \frac{1}{2} \times 20 \times 214$

$\Rightarrow S_{20} = 2140$

Find the value of n of the following arithmetic sequence such that its sum first exceeds 250.

12,22,32,42,52....

$a = 12$, $d = 10$, $S_n > 250$

$$S_n = \frac{1}{2}n(2a+(n-1)d)$$

$\Rightarrow$ $250 < \frac{1}{2}n(2\times 12+(n-1)\times 10)$

$\Rightarrow$ $250 < \frac{1}{2}n(14+10n)$

$\Rightarrow$ $500 < 14n+10n^2$

$\Rightarrow$ $10n^2 +14n -500 > 0$

solving quadratic

$10n^2 +14n -500 = 0$

$n = 6.4$ or $n = -7.8$ (both to 1dp)

The negative value can be discarded, since $n \in \mathbb{N}$

$\Rightarrow$ $n = 7$

An arithmetic sequence exists such that the sum of its first 6 terms is 93 and the sum of its first 9 terms is 207.

Find the sum of its first 4 terms.

$S_6 = 93$, $S_9 = 207$

$S_n = \dfrac{1}{2}n(2a + (n-1)d)$

$\Rightarrow 93 = \dfrac{1}{2} \times 6(2a + 5d)$ and $207 = \dfrac{1}{2} \times 9(2a + 8d)$

$\Rightarrow 186 = 6(2a + 5d)$ $\qquad\qquad 414 = 9(2a + 8d)$

$\Rightarrow 31 = 2a + 5d$ $\qquad\qquad\qquad 46 = 2a + 8d$

solve simultaneously

$15 = 3d$

$\Rightarrow d = 5$

substitute

$31 = 2a + 5d$

$\Rightarrow 31 = 2a + 25$

$\Rightarrow 6 = 2a$

$\Rightarrow a = 3$

$S_n = \dfrac{1}{2}n(2a + (n-1)d)$

$\Rightarrow S_n = \dfrac{1}{2}n(6 + 5(n-1))$

$\Rightarrow S_n = \dfrac{1}{2}n(1 + 5n)$

$\Rightarrow S_4 = \dfrac{1}{2} \times 4(1 + 5 \times 4)$

$\Rightarrow S_4 = 42$

When u$_n$ is given, the following formula can be used: -

$$S_n = \frac{1}{2}n(a+u_n)$$

since

$$S_n = \frac{1}{2}n(2a+(n-1)d)$$
$$\Rightarrow S_n = \frac{1}{2}n(a+a+(n-1)d)$$
but $u_n = a+(n-1)d$
$$\Rightarrow S_n = \frac{1}{2}n(a+u_n)$$

Example

Find the sum of the first six terms of the arithmetic sequence which starts 3,8,13... given u$_6$=28

$a = 3, \ u_6 = 28$

$$S_n = \frac{1}{2}n(a+u_n)$$
$$\Rightarrow S_6 = \frac{1}{2} \times 6(3+28)$$
$$\Rightarrow S_6 = 3 \times 31$$
$$\Rightarrow S_6 = 93$$

GEOMETRIC SEQUENCES

A geometric sequence is of the form

$$\frac{u_{n+3}}{u_{n+2}} = \frac{u_{n+2}}{u_{n+1}} = \frac{u_{n+1}}{u_n} = r$$

The common ratio, r, is a constant.

The nth term of a geometric sequence can be described

$$u_n = ar^{(n-1)}$$

where a = u_1, r = common ratio and n = n^{th} term

Examples

Find the 12th term of the geometric sequence 5, 20, 80...

$a = 5, \quad r = 4$

$\quad u_n = ar^{(n-1)}$

$\Rightarrow u_n = 5 \times 4^{(n-1)}$

$\Rightarrow u_{12} = 5 \times 4^{(12-1)}$

$\Rightarrow u_{12} = 5 \times 4^{11}$

$\Rightarrow u_{12} = 5 \times 4194304$

$\Rightarrow u_{12} = 20971520$

Given the geometric sequence   5, 20, 80,.....,

find the value of n for which $u_n = 327680$.

$a = 5, \quad r = 4, \quad u_n = 327680$

$\quad u_n = ar^{(n-1)}$

$\Rightarrow 327680 = 5 \times 4^{(n-1)}$

$\Rightarrow 65536 = 4^{(n-1)}$

$\Rightarrow \ln 65536 = \ln 4^{(n-1)}$

$\Rightarrow \ln 65536 = (n-1)\ln 4$

$\Rightarrow \dfrac{\ln 65536}{\ln 4} = (n-1)$

$\Rightarrow 8 = n - 1$

$\Rightarrow n = 9$

Given $u_5 = 1280$ and $u_8 = 81920$, find the geometric sequence.

$u_5 = 1280 \qquad u_8 = 81920$

$u_n = ar^{(n-1)}$

$\Rightarrow 1280 = ar^4 \quad \text{and} \quad 81920 = ar^7$

$\Rightarrow \dfrac{81920}{1280} = \dfrac{ar^7}{ar^4}$

$\Rightarrow 64 = r^3$

$\Rightarrow r = \sqrt[3]{64}$

$\Rightarrow r = 4$

substitute

$\quad 1280 = ar^4$

$\Rightarrow \quad 1280 = a \times 4^4$

$\Rightarrow \quad \dfrac{1280}{256} = a$

$\Rightarrow \quad a = 5$

$\quad u_n = ar^{(n-1)}$

$\Rightarrow \quad u_n = 5 \times 4^{(n-1)}$

SUM TO N TERMS OF A GEOMETRIC SEQUENCE

$$S_n = \frac{a(1-r^n)}{1-r}, \quad r \neq 1$$

Examples

Find the sum of the first seven terms of the geometric sequence 5, 20, 80...

$a = 5, \; r = 4$

$S_n = \dfrac{a(1-r^n)}{1-r}$

$\Rightarrow S_7 = \dfrac{5(1-4^7)}{1-4}$

$\Rightarrow S_7 = \dfrac{5(1-16384)}{-3}$

$\Rightarrow S_7 = \dfrac{5 \times (-16383)}{-3}$

$\Rightarrow S_7 = \dfrac{-81915}{-3}$

$\Rightarrow S_7 = 27305$

Given $S_5=1023$ and $S_8=65535$, find the geometric series.

$$S_5 = 1023 \qquad S_8 = 65535$$

$$S_n = \frac{a(1-r^n)}{1-r}$$

$\Rightarrow\ 1023 = \dfrac{a(1-r^5)}{1-r}\quad$ and $\quad 65535 = \dfrac{a(1-r^8)}{1-r}$

$\Rightarrow \dfrac{65535}{1023} = \dfrac{a(1-r^8)}{(1-r)} \times \dfrac{(1-r)}{a(1-r^5)}$

$\Rightarrow \dfrac{65535}{1023} = \dfrac{(1-r^8)}{(1-r^5)}$

$\Rightarrow 65535 - 65535r^5 = 1023 - 1023r^8$

$\Rightarrow 1023r^8 - 65535r^5 + 64512 = 0$

$\Rightarrow 341r^8 - 21845r^5 + 21504 = 0$

which factorises to

$(r-4)(r-1)(341r^6 + 1705r5 + 7161r^4 + 7140r^3 + 7056r^2 + 6720r + 5376) = 0$

with real solutions $r = 1$ or $r = 4$

discount $r = 1$

$$1023 = \frac{a(1-r^5)}{1-r}$$

$\Rightarrow 1023 = \dfrac{a(1-4^5)}{-3}$

$\Rightarrow \dfrac{-3069}{-1023} = a$

$\Rightarrow\quad a = 3$

series is $u_n = ar^{(n-1)}$

$3 + 12 + 48 + 192 + \ldots\ldots$

INFINITE SERIES

An **infinite series** has an infinite number of terms.

The sum of the first n terms, S_n, is called a **partial sum**.

If S_n tends to a limit as n tends to infinity, the limit is called the **sum to infinity** of the series.

ARITHMETIC SERIES

$$S_n = \frac{1}{2}n(2a + (n-1)d)$$
$$= \frac{1}{2}n(2a + nd - d)$$
$$= an + \frac{1}{2}n^2 d - \frac{1}{2}nd$$
$$= \frac{1}{2}n^2 d + n(a - \frac{1}{2}d)$$
$$= \frac{1}{2}n^2 d + n(a - \frac{1}{2}d) \times \frac{n}{n}$$
$$= \frac{1}{2}n^2 d + \frac{n^2}{n}(a - \frac{1}{2}d)$$
$$= n^2 \left(\frac{1}{2}d + \frac{1}{n}(a - \frac{1}{2}d) \right)$$

As n tends to infinity, S_n tends to $\pm \infty$

The sum to infinity for an arithmetic series is undefined.

GEOMETRIC SERIES

$$S_n = \frac{a(1-r^n)}{1-r}, \quad r \neq 1$$

When r > 1, r^n tends to infinity as n tends to infinity.
When r < 1, r^n tends to zero as n tends to infinity.

The sum to infinity for a geometric series is undefined when $|r| > 1$

The sum to infinity for a geometric series when $|r| < 1$ is

$$S_\infty = \frac{a}{1-r}$$

Examples

Find the sum to infinity for the series   96 + 48 + 24...... if it exists.

$a = 96, \quad r = 0.5$

$|r| < 1 \Rightarrow S_\infty$ exists

$$S_\infty = \frac{a}{1-r}$$

$$\Rightarrow S_\infty = \frac{96}{1-0.5}$$

$$\Rightarrow S_\infty = \frac{96}{0.5}$$

$$\Rightarrow S_\infty = 192$$

Express the recurring decimal 0.242424.... as a vulgar fraction.

$a = 0.24, \quad r = 1/100$

$|r| < 1 \Rightarrow S_\infty$ exists

$S_\infty = \dfrac{a}{1-r}$

$\Rightarrow S_\infty = \dfrac{0.24}{1-0.01}$

$\Rightarrow S_\infty = \dfrac{0.24}{0.99}$

$\Rightarrow S_\infty = \dfrac{24}{99}$

$\Rightarrow S_\infty = \dfrac{8}{33}$

Given that 12 and 6 are two adjacent terms of an infinite geometric series with a sum to infinity of 192,

a) find the first term; and
b) find the partial sum S_6

a)

$u_a = 12, \quad u_{a+1} = 6, \quad r = 6/12 = 1/2$

$S_\infty = 192$

$S_\infty = \dfrac{a}{1-r}$

$\Rightarrow 192 = \dfrac{a}{1-0.5}$

$\Rightarrow 192 = \dfrac{a}{0.5}$

$\Rightarrow a = 384$

b)

$a = 384, \quad r = 0.5$

$S_6 = \dfrac{a(1-r^n)}{1-r}$

$\Rightarrow S_6 = \dfrac{384(1-0.5^6)}{1-0.5}$

$\Rightarrow S_6 = \dfrac{384(1-0.5^6)}{0.5}$

$\Rightarrow S_6 = 756$

SIGMA NOTATION - RULES

$$\sum_{r=1}^{n} f(r) = f(1) + f(2) + f(3) + f(4) + f(5) + \ldots + f(n)$$

Constant multiplier

$$\sum_{r=1}^{n} kf(r) = kf(1) + kf(2) + kf(3) + kf(4) + kf(5) + \ldots + kf(n)$$
$$= kf(1) + kf(2) + kf(3) + kf(4) + kf + \ldots + kf(n)$$
$$= k(f(1) + f(2) + f(3) + f(4) + f + \ldots + f(n))$$
$$= k(f(1) + f(2) + f(3) + f(4) + f + \ldots + f(n))$$
$$= k\sum_{r=1}^{n} f(r)$$

Adding series

$$\sum_{r=1}^{n} \bigl(f(r) + g(r)\bigr) = f(1) + g(1) + f(2) + g(2) + f(3) + g(3) + \ldots + f(n) + g(n)$$
$$= f(1) + f(2) + f(3) + \ldots f(n). + g(1) + g(2) + g(3) + \ldots + g(n)$$
$$= \sum_{r=1}^{n} f(r) + \sum_{r=1}^{n} g(r)$$

SUM OF FIRST N NATURAL NUMBERS

$$\sum_{r=1}^{n} r = 1 + 2 + 3 + 4 + 5 + \ldots + n$$

written backwards

$$\sum_{r=1}^{n} r = n + (n-1) + (n-2) + \ldots + 3 + 2 + 1$$

added

$$2\sum_{r=1}^{n} r = (1+n) + (2+n-1) + (3+n-2) + \ldots + (n+1)$$

$$= (n+1) + (n+1) + (n+1) + \ldots + (n+1)$$

$$\therefore \quad 2\sum_{r=1}^{n} r = n(n+1)$$

$$\Rightarrow \quad \sum_{r=1}^{n} r = \frac{n(n+1)}{2} \qquad \boxed{\sum_{r=1}^{n} r = \frac{n(n+1)}{2}}$$

$$\sum_{r=1}^{n}(ar+b) = a+b+2a+b+3a+b+\ldots+na+b$$
$$= a+2a+3a+\ldots+na+b+b+b+\ldots+b$$
$$= a(1+2+3+\ldots+n)+b(1+1+1\ldots+1)$$
$$= a\sum_{r=1}^{n}r + b\sum_{r=1}^{n}1$$
$$= a \times \frac{1}{2}n(a+u_n)+b \times \frac{1}{2}n(a+u_n)$$
$$= \frac{1}{2}an(1+n)+b \times \frac{1}{2}n(1+1)$$
$$= \frac{1}{2}an(1+n)+bn$$

Example

Find the value of $\sum_{r=1}^{n}6r+5$

$$\sum_{r=1}^{n}6r+5 = (6\times 1+5)+(6\times 2+5)+(6\times 3+5)+\ldots+(6\times n+5)$$
$$= 11+17+23+\ldots+(6\times n+5)$$

COMMON SERIES – SIGMA NOTATION

Even numbers
$$\sum_{r=1}^{n} 2r = 2+4+6+8+10+\ldots+n$$

Odd numbers
$$\sum_{r=1}^{n} 2r-1 = 1+3+5+7+\ldots+n$$
or
$$\sum_{r=1}^{n} 2r+1 = 3+5+7+\ldots+n$$

$$\sum_{r=1}^{n} r = 1+2+3+4+5+\ldots+n$$

$$\sum_{r=1}^{n} r^2 = 1+4+9+16+25+\ldots+n^2$$

$$\sum_{r=1}^{n} x^r = 1+x+x^2+x^3+x^4+x^5+\ldots+x^n$$

$$\frac{1}{1-x} = \sum_{r=0}^{\infty} x^r = 1+x+x^2+x^3+x^4+x^5+\ldots+x^{n-1} \quad, \; -1<x<1$$

$$\sum_{r=1}^{n} (-1)^r = (-1)^1+(-1)^2+(-1)^3+(-1)^4+(-1)^5+\ldots+(-1)^n$$
$$= -1+1-1+1-1+\ldots+(-1)^n$$

$$\sum_{r=1}^{n} (-1)^{r+1} = (-1)^{1+1}+(-1)^{2+1}+(-1)^{3+1}+(-1)^{4+1}+(-1)^{5+1}+\ldots+(-1)^{n+1}$$
$$= 1-1+1-1+1+\ldots+(-1)^{n+1}$$

POWER SERIES

$$\sum_{n=0}^{\infty} a_n x^n = a_0 + a_1 x + a_2 x^2 + a_3 x^3 + a_4 x^4 + \ldots$$

where $a_0, a_1, a_2 \ldots a_n$ are constants.

The series always converges when $x = 0$
It will possibly converge for other values of x.
A series cannot be convergent unless its terms tend to zero.

$$\operatorname*{Lim}_{n \to \infty} u_n = 0$$

D'ALEMBERT'S RATIO TEST

For a series of positive terms

$$\lim_{n \to \infty} \left(\frac{u_{n+1}}{u_n} \right) < 1 \quad \Rightarrow \text{ series converges}$$

$$\lim_{n \to \infty} \left(\frac{u_{n+1}}{u_n} \right) > 1 \quad \Rightarrow \text{ series diverges}$$

$$\lim_{n \to \infty} \left(\frac{u_{n+1}}{u_n} \right) = 1 \quad \Rightarrow \text{ indeterminate}$$

Example

Use d'Alembert's ratio test to test for convergence of the following series :-

$$1 + \frac{3}{2} + \frac{5}{4} + \frac{7}{8} + \ldots$$

$$1 + \frac{3}{2} + \frac{5}{4} + \frac{7}{8} + \ldots = \frac{1}{1} + \frac{3}{2} + \frac{5}{2^2} + \frac{7}{2^3} + \ldots$$

$$\therefore u_n = \frac{2n-1}{2^{(n-1)}} \quad \text{and} \quad u_{n+1} = \frac{2(n+1)-1}{2^{(n+1-1)}} = \frac{2n+1}{2^n}$$

$$\frac{u_{n+1}}{u_n} = \frac{2n+1}{2^n} \times \frac{2^{(n-1)}}{2n-1}$$

$$= \frac{2n+1}{2n-1} \times \frac{1}{2}$$

$$\lim_{n \to \infty} \left(\frac{u_{n+1}}{u_n} \right) = \lim_{n \to \infty} \frac{2n+1}{2n-1} \times \frac{1}{2}$$

$$= \lim_{n \to \infty} \frac{2+1/n}{2-1/n} \times \frac{1}{2}$$

$$= \frac{2+0}{2-0} \times \frac{1}{2}$$

$$= 1 \times \frac{1}{2}$$

$$= \frac{1}{2}$$

$\lim_{n \to \infty} \left(\frac{u_{n+1}}{u_n} \right) < 1$ so series converges

ABSOLUTE CONVERGENCE

If $\sum |u_n|$ is convergent, $\sum u_n$ is absolutely convergent.

if $\sum |u_n|$ is divergent and $\sum u_n$ converges,
then $\sum u_n$ is conditionally convergent.

$$\lim_{n \to \infty} \left| \frac{a_{n+1} x}{a_n} \right| < 1 \quad \Rightarrow \quad \text{series converges absolutely}$$

$$\lim_{n \to \infty} \left| \frac{a_{n+1} x}{a_n} \right| > 1 \quad \Rightarrow \quad \text{series diverges or converges conditionally}$$

$$\lim_{n \to \infty} \left| \frac{a_{n+1} x}{a_n} \right| = 1 \quad \Rightarrow \quad \text{indeterminate}$$

Example

Find the sum to infinity of the series
$4 + 7x + 10x^2 + 13x^3 + ...$
and the region of valid values of x.

Let $S_n = 4 + 7x + 10x^2 + 13x^3 + ... + (3n+1)x^{n-1}$

multiply by x

$x S_n = 4x + 7x^2 + 10x^3 + 13x^4 + ... + (3n+1)x^n$

subtract

$S_n(1-x) = 4 + 3x + 3x^2 + 3x^3 + ... + 3x^{n-1} - (3n+1)x^n$

$\qquad = 4 + 3(x + x^2 + x^3 + ... + x^{n-1}) - (3n+1)x^n$

$\qquad = 4 + 3\left(\dfrac{x(1-x^{n-1})}{1-x}\right) - (3n+1)x^n$

if $-1 < x < 1$ then as $n \to \infty$, $x^n \to 0, x^{n-1} \to 0$

$\Rightarrow S_\infty(1-x) = 4 + 3\left(\dfrac{x}{1-x}\right)$

$\Rightarrow S_\infty(1-x) = \dfrac{4(1-x) + 3x}{1-x}$

$\Rightarrow S_\infty(1-x) = \dfrac{4-x}{1-x}$

$\Rightarrow S_\infty = \dfrac{4-x}{(1-x)^2}$ for $-1 < x < 1$

FIBONACCI SERIES

If F_n is the nth Fibonacci number

$$\lim_{n \to \infty} \frac{F_{n+1}}{F_n} = \frac{1+\sqrt{5}}{2}$$

$$\lim_{n \to \infty} \left| \frac{F_{n+1} x^{n+1}}{F_n x^n} \right| < 1 \Rightarrow \frac{1+\sqrt{5}}{2} |x| < 1$$

converges when

$$|x| < \frac{2}{1+\sqrt{5}}$$

CENTRE OF CONVERGENCE

The power series can be written

$$\sum_{n=0}^{\infty} a_n (x-c)^n$$
$$= \lim_{n \to \infty} a_0 + a_1(x-c) + a_2(x-c)^2 + a_3(x-c)^3 + \ldots + a_n(x-c)^n$$

where c is the centre of convergence.
This is the middle of the interval of convergence,
the interval for which the limit exists.
The radius of convergence is called R.

If R = ∞ , the series converges for all x.
Otherwise,

the series converges for $|x-c| < R$

and diverges for $|x-c| > R$

$$f(x) = a_0 + a_1(x-c) + a_2(x-c)^2 + a_3(x-c)^3 + ...$$

To find a_0, set x = c , since the remaining terms will become zero.

differentiate

$$f'(x) = a_1 + 2a_2(x-c) + 3a_3(x-c)^2 + 4a_4(x-c)^3 + ...$$

To find a_1, evaluate f'(c)
go again

$$f''(x) = 2a_2 + 3.2a_3(x-c) + 4.3a_4(x-c)^2 + ...$$
$$f''(c) = 2a_2$$
$$\Rightarrow a_2 = \frac{f''(c)}{2}$$
$$\Rightarrow a_2 = \frac{f''(c)}{2!}$$

Continuing gives the following

$$f'''(x) = 3.2a_3 + 4.3.2a_4(x-c) + 5.4.3a_5(x-c)^2 + \ldots$$

$$f'''(c) = 3.2a_3$$

$$\Rightarrow a_3 = \frac{f'''(c)}{3!}$$

and

$$f^{iv}(x) = 4.3.2a_4 + 5.4.3.2a_5(x-c) + \ldots$$

$$f^{iv}(c) = 4.3.2a_4$$

$$\Rightarrow a_4 = \frac{f^{iv}(c)}{4!}$$

so that $\boxed{a_n = \frac{f^n(c)}{n!}}$

Substituting back into the original series

$$f(x) = a_0 + a_1(x-c) + a_2(x-c)^2 + a_3(x-c)^3 + \ldots$$

$$= f(c) + f'(c)(x-c) + \frac{f''(c)(x-c)^2}{2} + \frac{f'''(c)(x-c)^3}{3.2} + \ldots$$

$$= f(c) + f'(c)(x-c) + \frac{f''(c)(x-c)^2}{2!} + \frac{f'''(c)(x-c)^3}{3!} + \ldots$$

$$\boxed{f(x) = \sum_{n=0}^{\infty} \frac{f^n(c)}{n!}(x-c)^n}$$

$$= f(c) + f'(c)(x-c) + \frac{f''(c)(x-c)^2}{2!} + \frac{f'''(c)(x-c)^3}{3!} + \ldots$$

In the interval $(-R+c, R+c)$

TAYLOR'S SERIES

$$f(x+h) = \sum_{n=0}^{\infty} \frac{h^n}{n!} f^{(n)}(x)$$
$$= f(x) + hf'(x) + \frac{h^2 f''(x)}{2!} + \frac{h^3 f'''(x)}{3!} + \ldots$$

MACLAURIN'S SERIES

A power series centred at c may be written

$$f(x) = a_0 + a_1(x-c) + a_2(x-c)^2 + a_3(x-c)^3 + \ldots$$

When c = 0, the power series is

$$\sum_{n=0}^{\infty} a_n x^n = a_0 + a_1 x + a_2 x^2 + a_3 x^3 + a_4 x^4 + \ldots$$

let $f(x) = a_0 + a_1 x + a_2 x^2 + a_3 x^3 + a_4 x^4 + \ldots$

Differentiate and set x to zero:-

$$f'(x) = b + 2cx + 3dx^2 + 4ex^3 + 5fx^4 + 6g^5 x + \ldots$$
$$f'(0) = b$$

And again,
$$f''(x) = 2c + 2.3dx + 3.4ex^2 + 4.5fex^3 + 5.6gx^4 + \ldots$$

$f''(0) = 2c$

$\Rightarrow c = \dfrac{f''(0)}{2}$

and again,

$f'''(x) = 2.3d + 2.3.4ex + 3.4.5fex^2 + 4.5.6gx^3 + \ldots$

$f'''(0) = 2.3d$

$\Rightarrow d = \dfrac{f'''(0)}{2.3}$

and again,

$f^{iv}(x) = 2.3.4e + 2.3.4.5fex + 3.4.5.6gx^2 + \ldots$

$f^{iv}(0) = 2.3.4e$

$\Rightarrow d = \dfrac{f^{iv}(0)}{2.3.4}$

etc.

Substituting back into the original series

$f(x) = a + bx + cx^2 + dx^3 + ex^4 + fx^5 + gx^6 + \ldots$

$= f(0) + f'(0)x + \dfrac{f''(0)}{2}x^2 + \dfrac{f'''(0)}{2.3}x^3 + \dfrac{f^{iv}(0)}{2.3.4}x^4 + \ldots$

$= f(0) + f'(0)x + \dfrac{f''(0)x^2}{2} + \dfrac{f'''(0)x^3}{3.2} + \dfrac{f^{iv}(0)x^4}{4.3.2} + \ldots$

Giving Maclaurin's series,

$$f(x) = \frac{f(0)}{0!} + \frac{f'(0)x}{1!} + \frac{f''(0)x^2}{2!} + \frac{f'''(0)x^3}{3!} + \frac{f^{iv}(0)x^4}{4!} + \ldots\ldots$$

$$f(x) = \sum_{n=0}^{\infty} \frac{f^{(n)}(0)}{n!} x^n$$

which has valid values for x when

$$\lim_{n \to \infty} \left| \frac{u_{n+1} x}{u_n} \right| < 1$$

$$\lim_{n \to \infty} \left| \frac{f^{(n+1)}(0)}{(n+1)!} x^{n+1} \times \frac{n!}{f^{(n)}(0) x^n} \right| < 1$$

$$\lim_{n \to \infty} \left| \frac{x f^{(n+1)}(0)}{(n+1)} \times \frac{1}{f^{(n)}(0)} \right| < 1$$

$$\lim_{n \to \infty} \left| \frac{x f^{(n+1)}(0)}{(n+1) f^{(n)}(0)} \right| < 1$$

Example

Find the Maclaurin expansion of $f(x) = (1+x)^n$,

Differentiate

$f(x) = (1+x)^n$

$f'(x) = n(1+x)^{n-1}$

$f''(x) = n(n-1)(1+x)^{n-2}$

$f'''(x) = n(n-1)(n-2)(1+x)^{n-3}$

$f^{iv}(x) = n(n-1)(n-2)(n-3)(1+x)^{n-4}$

etc

set $x = 0$

$f(0) = 1$

$f'(0) = n$

$f''(0) = n(n-1)$

$f'''(0) = n(n-1)(n-2)$

$f^{iv}(0) = n(n-1)(n-2)(n-3)$

etc

$$f(x) = \frac{f(0)}{0!} + \frac{f'(0)x}{1!} + \frac{f''(0)x^2}{2!} + \frac{f'''(0)x^3}{3!} + \frac{f^{iv}(0)x^4}{4!} + \ldots\ldots$$

$$(1+x)^n = 1 + nx + \frac{n(n-1)x^2}{2!} + \frac{n(n-1)(n-2)x^3}{3!} + \frac{n(n-1)(n-2)(n-3)x^4}{4!} + \ldots.$$

which is the Binomial series.

Find the Maclaurin expansion of
f(x)=(1-x)n

$f(x) = (1-x)^n = (1+(-x))^n$

$(1+x)^n = 1 + nx + \dfrac{n(n-1)x^2}{2!} + \dfrac{n(n-1)(n-2)x^3}{3!} + \dfrac{n(n-1)(n-2)(n-3)x^4}{4!} + \ldots$

so

$(1-x)^n = 1 - nx + \dfrac{n(n-1)x^2}{2!} - \dfrac{n(n-1)(n-2)x^3}{3!} + \dfrac{n(n-1)(n-2)(n-3)x^4}{4!} - \ldots + \ldots$

Use the Maclaurin Series to find a series for
f(x) = sinx

$f(x) = \sin x = a + bx + cx^2 + dx^3 + ex^4 + fx^5 + gx^6 + \ldots$

$f'(x) = \cos x = b + 2cx + 3dx^2 + 4ex^3 + 5fx^4 + 6gx^5 + \ldots$
$f''(x) = -\sin x = 2c + 2.3dx + 3.4ex^2 + 4.5fex^3 + 5.6gx^4 + \ldots$
$f'''(x) = -\cos x = 2.3d + 2.3.4ex + 3.4.5fex^2 + 4.5.6gx^3 + \ldots$
$f^{iv}(x) = \sin x = 2.3.4e + 2.3.4.5fex + 3.4.5.6gx^2 + \ldots$

when $x = 0$
$\sin 0 = a \Rightarrow a = 0$
$f'(0) = \cos 0 = b \Rightarrow b = 1$
$f''(0) = -\sin 0 = c \Rightarrow c = 0$
$f'''(0) = -\cos 0 = d \Rightarrow d = -1$

$f(x) = \dfrac{f(0)}{0!} + \dfrac{f'(0)x}{1!} + \dfrac{f''(0)x^2}{2!} + \dfrac{f'''(0)x^3}{3!} + \dfrac{f^{iv}(0)x^4}{4!} + \ldots$

$\sin x = \dfrac{0}{0!} + \dfrac{1.x}{1!} + \dfrac{0.x^2}{2!} + \dfrac{-1.x^3}{3!} + \dfrac{0.x^4}{4!} + \ldots$

$= 0 + x - \dfrac{x^3}{3!} + \dfrac{x^5}{5!} - \dfrac{x^7}{7!} + \ldots$

Use the Maclaurin Series to find a series for
$f(x) = (1+x)^{1/4}$

Differentiate   set $x = 0$

$f(x) = (1+x)^{1/4}$ $f(0) = 1$

$f'(x) = \dfrac{1}{4}(1+x)^{-3/4}$ $f'(0) = \dfrac{1}{4}$

$f''(x) = \dfrac{1}{4} \cdot \dfrac{(-3)}{4}(1+x)^{-7/4}$ $f''(0) = \dfrac{1}{4} \cdot \dfrac{(-3)}{4} = \dfrac{-3}{16}$

$f'''(x) = \dfrac{1}{4} \cdot \dfrac{(-3)}{4} \cdot \dfrac{(-7)}{4}(1+x)^{-11/4}$ $f'''(0) = \dfrac{1}{4} \cdot \dfrac{(-3)}{4} \cdot \dfrac{(-7)}{4} = \dfrac{21}{64}$

$f^{iv}(x) = \dfrac{1}{4} \cdot \dfrac{(-3)}{4} \cdot \dfrac{(-7)}{4} \cdot \dfrac{(-11)}{4}(1+x)^{-15/4}$ $f^{iv}(0) = \dfrac{1}{4} \cdot \dfrac{(-3)}{4} \cdot \dfrac{(-7)}{4} \cdot \dfrac{(-11)}{4} = \dfrac{-231}{256}$

etc

$f(x) = \dfrac{f(0)}{0!} + \dfrac{f'(0)x}{1!} + \dfrac{f''(0)x^2}{2!} + \dfrac{f'''(0)x^3}{3!} + \dfrac{f^{iv}(0)x^4}{4!} + \ldots$

$(1+x)^{1/4} = 1 + \dfrac{1}{4}x - \dfrac{3x^2}{16 \cdot 2!} + \dfrac{21x^3}{64 \cdot 3!} - \dfrac{231x^4}{256 \cdot 4!} + \ldots$

$= 1 + \dfrac{1}{4}x - \dfrac{3x^2}{32} + \dfrac{21x^3}{384} - \dfrac{231x^4}{6144} + \ldots$

$= 1 + \dfrac{1}{4}x - \dfrac{3x^2}{32} + \dfrac{7x^3}{128} - \dfrac{77x^4}{2048} + \ldots$

Find the Maclaurin expansion of
f(x)=ex

$$f(x) = e^x = a + bx + cx^2 + dx^3 + ex^4 + fx^5 + gx^6 + ...$$

Differentiate

$$f'(x) = e^x = b + 2cx + 3dx^2 + 4ex^3 + 5fx^4 + 6g^5x + ...$$
$$f''(x) = e^x = 2c + 2.3dx + 3.4ex^2 + 4.5fex^3 + 5.6gx^4 + ...$$
$$f'''(x) = e^x = 2.3d + 2.3.4ex + 3.4.5fex^2 + 4.5.6gx^3 + ...$$
$$f^{iv}(x) = e^x = 2.3.4e + 2.3.4.5fex + 3.4.5.6gx^2 + ...$$

etc

set $x = 0$

$$f(0) = e^0 = a \Rightarrow a = 1$$
$$f'(0) = e^0 = b \Rightarrow b = 1$$
$$f''(0) = e^0 = 2c \Rightarrow c = 1/2$$
$$f'''(0) = e^0 = 6d \Rightarrow d = 1/6$$
$$f^{iv}(0) = e^0 = 12e \Rightarrow e = 1/12$$

etc

$$f(x) = a + bx + cx^2 + dx^3 + ex^4 + fx^5 + gx^6 + ...$$
$$= f(0) + f'(0)x + \frac{f''(0)}{2}x^2 + \frac{f'''(0)}{2.3}x^3 + \frac{f^{iv}(0)}{2.3.4}x^4 + ...$$

$$e^x = 1 + 1.x + \frac{1.x^2}{2} + \frac{1.x^3}{6} + \frac{1.x^4}{12} + ...$$

$$= 1 + x + \frac{x^2}{2} + \frac{x^3}{6} + \frac{x^4}{12} + ...$$

$$= 1 + \frac{x}{1!} + \frac{x^2}{2!} + \frac{x^3}{3!} + \frac{x^4}{4!} + ...$$

STANDARD SERIES

$$e^x = \sum_{n=0}^{\infty} \frac{x^n}{n!} = 1 + \frac{x}{1!} + \frac{x^2}{2!} + \frac{x^3}{3!} + \frac{x^4}{4!} + \ldots\ldots$$

$$e^{ax} = \sum_{n=0}^{\infty} \frac{(ax)^n}{n!} = \sum_{n=0}^{\infty} a^n \frac{x^n}{n!} = 1 + \frac{(ax)}{1!} + \frac{(ax)^2}{2!} + \frac{(ax)^3}{3!} + \frac{(ax)^4}{4!} + \ldots\ldots$$

$$e^{ax+b} = e^{ax} e^b = \sum_{n=0}^{\infty} a^n e^b \frac{x^n}{n!} = \frac{e^b (ax)}{1!} + \frac{e^b (ax)^2}{2!} + \frac{e^b (ax)^3}{3!} + \ldots\ldots$$

$$\ln(1+x) = \sum_{n=0}^{\infty} (-1)^{n+1} \frac{x^n}{n} = x - \frac{x^2}{2} + \frac{x^3}{3} - \frac{x^4}{4} + \frac{x^5}{5} - \ldots + \ldots\ldots$$

$$\ln(1-x) = -\sum_{n=0}^{\infty} \frac{x^n}{n} = -x - \frac{x^2}{2} - \frac{x^3}{3} - \frac{x^4}{4} - \frac{x^5}{5} - \ldots\ldots$$

provided $-1 < x < 1$

$$\sin x = \sum_{n=0}^{\infty} \frac{(-1)^n}{(2n+1)!} x^{2n+1} = x - \frac{x^3}{3!} + \frac{x^5}{5!} - \frac{x^7}{7!} + \frac{x^9}{9!} - \ldots + \ldots\ldots$$

$$\cos x = \sum_{n=0}^{\infty} \frac{(-1)^n}{(2n)!} x^{2n} = 1 - \frac{x^2}{2!} + \frac{x^4}{4!} - \frac{x^6}{6!} + \frac{x^8}{8!} - \frac{x^{10}}{10!} + \ldots$$

$$(a+x)^r = a^r + ra^{r-1}x + \frac{r(r-1)}{2!} a^{r-2} x^2 + \frac{r(r-1)(r-2)}{3!} a^{r-3} x^3 + \ldots$$

provided $-a < x < a$

Examples

Find the first four terms of a power series for cos3x

$$\cos x = \sum_{n=0}^{\infty} \frac{(-1)^n}{(2n)!}(3x)^{2n} = 1 - \frac{(3x)^2}{2} + \frac{(3x)^4}{4} - \frac{(3x)^6}{6} + \ldots$$

$$= 1 - \frac{9x^2}{2} + \frac{81x^4}{4} - \frac{729x^6}{6} + \ldots$$

$$= 1 - \frac{9x^2}{2} + \frac{81x^4}{4} - \frac{243x^6}{3} + \ldots$$

Express the following as a power series in x:-

$$\sin\left(3x + \frac{\pi}{2}\right)$$

$$\sin\left(3x + \frac{\pi}{4}\right) = \sin 3x \cos\frac{\pi}{4} + \cos 3x \sin\frac{\pi}{4}$$

$$= \frac{1}{\sqrt{2}} \sin 3x + \frac{1}{\sqrt{2}} \cos 3x$$

$$= \frac{1}{\sqrt{2}} \sum_{n=0}^{\infty} \frac{(-1)^n}{(2n+1)!}(3x)^{2n+1} + \frac{1}{\sqrt{2}} \sum_{n=0}^{\infty} \frac{(-1)^n}{(2n)!}(3x)^{2n}$$

$$= \frac{1}{\sqrt{2}} \sum_{n=0}^{\infty} \frac{(-1)^n}{(2n+1)!}(3x)^{2n+1} + \frac{(-1)^n}{(2n)!}(3x)^{2n}$$

$$= \frac{1}{\sqrt{2}} \sum_{n=0}^{\infty} \frac{(-1)^n}{(2n)!}(3x)^{2n} + \frac{(-1)^n}{(2n+1)!}(3x)^{2n+1}$$

$$= \frac{1}{\sqrt{2}} \sum_{n=0}^{\infty} \frac{(-1)^n}{(2n)!}(3x)^{2n}\left(1 + \frac{3x}{2n+1}\right)$$

ITERATION

An iterative sequence is one generated by the recurrence relation $x_{n+1} = F(x_n)$.

The starting value is x_0 and each term is called an iterate.
A fixed point, or convergent, occurs when $x_n = F(x_n)$.

Example
Given $x_{n+1} = 4 - 3x_n$ and starting point $x_0 = 5$,
a) calculate the first five iterates.
b) find any fixed points which exist

a)
$x_0 = 5$
$x_1 = 4 - 3x_0 = 4 - 15 = -11$
$x_2 = 4 - 3x_1 = 4 + 33 = 37$
$x_3 = 4 - 3x_2 = 4 - 111 = -107$
$x_4 = 4 - 3x_3 = 4 + 321 = 325$
$x_5 = 4 - 3x_4 = 4 - 975 = -971$

b) For fixed points, $x_n = F(x_n)$
so $x_{n+1} = x_n = x$

$x = 4 - 3x$

so $4x = 4$

$x = 1$

$x = 1$ is a fixed point

NEWTON RAPHSON ITERATION

$$x_{n+1} = x_n - \frac{f(x_n)}{f'(x_n)}$$

GRAPHICAL METHODS

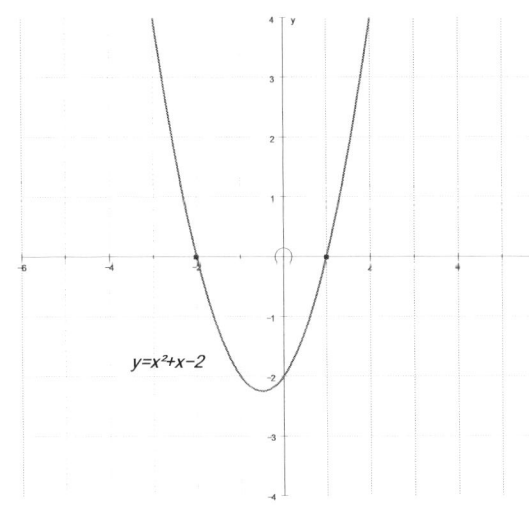

Consider the function $f(x) = x^2 + x - 2$ which is graphed left.

From the graph, $f(x) = 0$ at the points $(-2, 0)$ and $(1, 0)$

Consider now re-arranging $x^2 + x - 2 = 0$
to $x = 2 - x^2$

x is now in the form $x = F(x)$, where $F(x) = 2 - x^2$

Plotting $y = 2 - x^2$ gives the graph

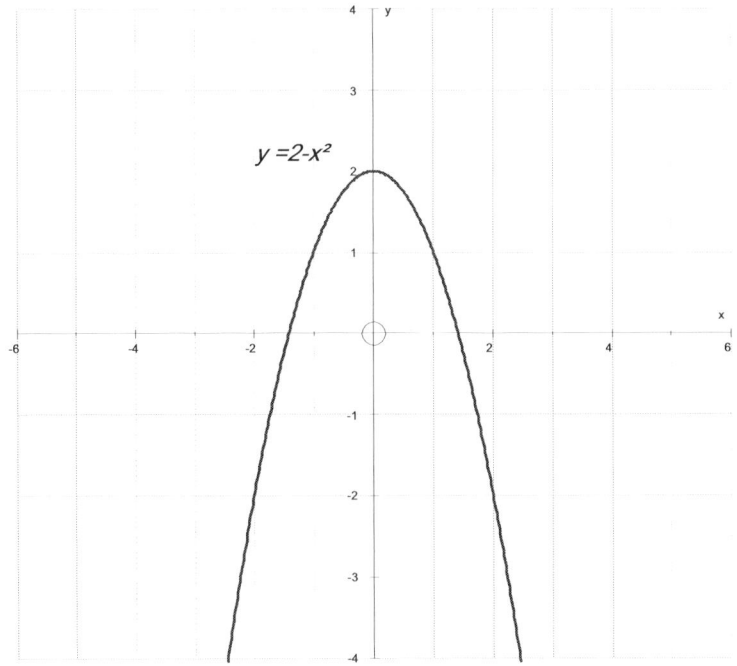

but $y = 2 - x^2$ and $x = 2 - x^2$
So $y = x = 2 - x^2$ i.e. $y = x = F(x)$

The intersection of y = x and y = 2 − x²

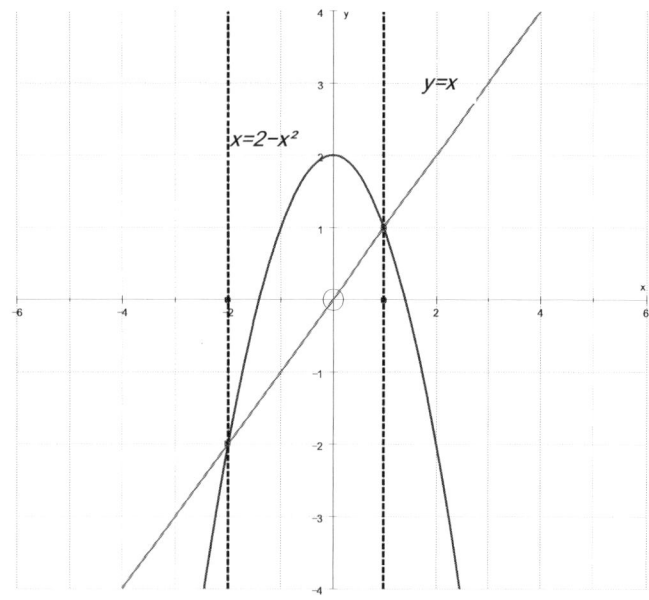

gives the points (−2,0) and (1, 0)

Which is the same as f(x) = 0

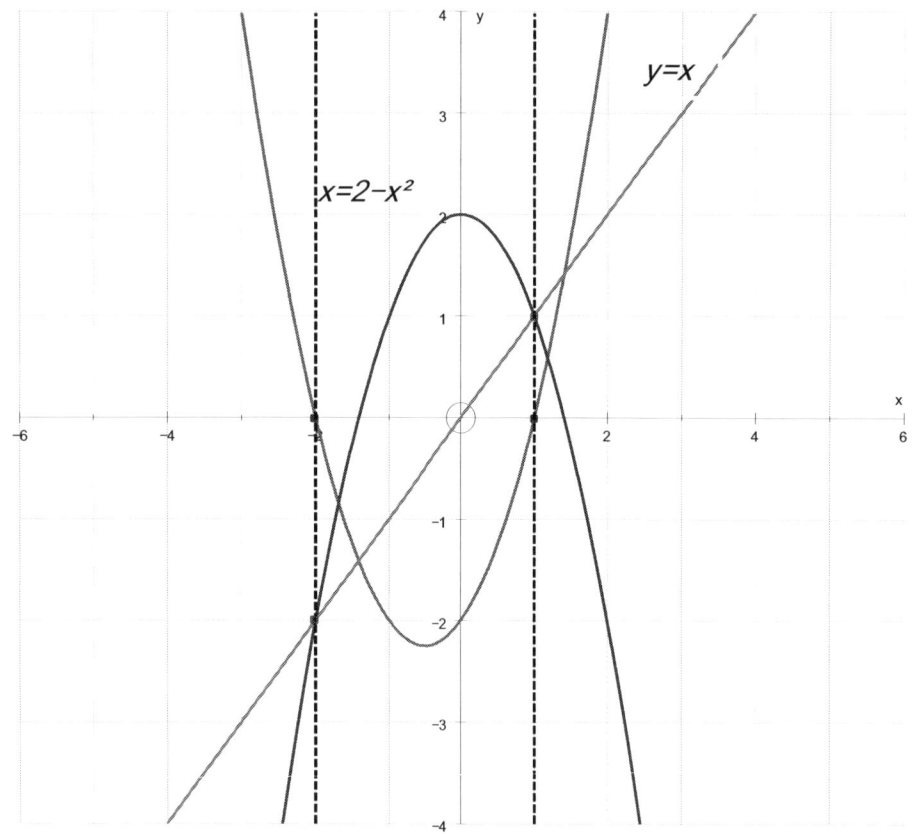

An iterative sequence for $x = 2 - x^2$ would be of the form

$x_{n+1} = 2 - x_n^2$ for fixed points, $x_n = F(x_n)$ so $x_{n+1} = x_n = x$

$x_{n+1} = 2 - x_n^2$ has fixed points,

$x = 2 - x^2$
$\Rightarrow x^2 + x - 2 = 0$
$\Rightarrow (x+2)(x-1) = 0$
$\Rightarrow x = -2$ or $x = 1$

which are the intersections found above.

So, if an equation $f(x) = 0$ can be rearranged to the form $x = F(x)$, finding any fixed points of the iterative sequence $x_{n+1} = F(x_n)$ may lead to solutions for $f(x)=0$.

Example

Given that $x^3 - 4x - 5 = 0$ can be written
as $x = (4x + 5)^{1/3}$ and using a starting value of 2:-
a) Write down a simple iterative formula $x_{n+1} = F(x_n)$
b) Find, correct to 4dp, a root between 1 and 3

a) $x = (4x + 5)^{1/3}$ so $x_{n+1} = (4x_n + 5)^{1/3}$

b) $x_{n+1} = (4x_n + 5)^{1/3}$

	x_n	$4x_n$	$4x_n + 5$	$(4x_n + 5)^{1/3}$
x_0	2	8	13	2.351335
x_1	2.35133	9.405339	14.405339	2.433181
x_2	2.43318	9.732726	14.732726	2.451476
x_3	2.45148	9.805905	14.805905	2.455529
x_4	2.45553	9.822114	14.822114	2.456424
x_5	2.45642	9.825697	14.825697	2.456622
x_6	2.45662	9.826489	14.826489	2.456666
x_7	2.45667	9.826664	14.826664	2.456676
x_8	2.45668	9.826702	14.826702	2.456678
x_9	2.45668	9.826711	14.826711	2.456678
x_{10}	2.45668	9.826713	14.826713	2.456678
x_{11}	2.45668	9.826713	14.826713	2.456678
x_{12}	2.45668	9.826713	14.826713	2.456678
x_{13}	2.45668	9.826713	14.826713	2.456678
x_{14}	2.45668	9.826713	14.826713	2.456678

$x^3 - 4x - 5 = 0$ has a root 2.4567 (4dp)

Below is a graph showing the same iterative formula, this time with a starting value of $x_0 = -4$

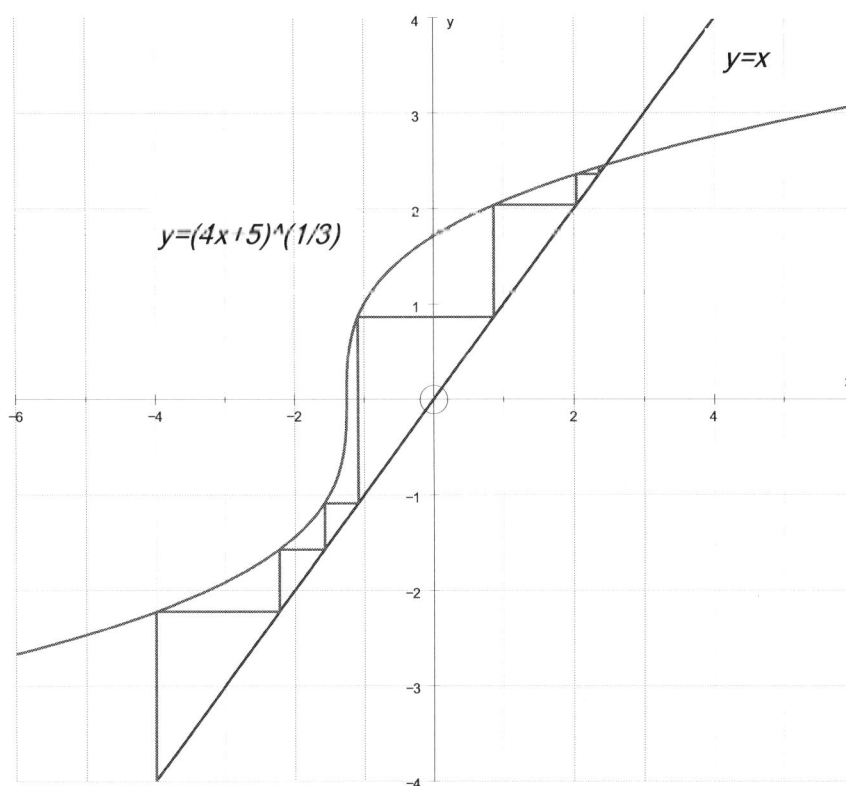

$$x_{n+1}=(4x_n+5)^{1/3}$$

	x_n	$4x_n$	$4x_n+5$	$(4x_n+5)^{1/3}$
x_0	−4	−16	−11	−2.22398
x_1	−2.22398	−8.89592	−3.89592	−1.57351
x_2	−1.57351	−6.29405	−1.294047	−1.08972
x_3	−1.08972	−4.3589	0.6411018	0.862268
x_4	0.86227	3.449072	8.4490725	2.036744
x_5	2.03674	8.146974	13.146974	2.360163
x_6	2.36016	9.440651	14.440651	2.435168
x_7	2.43517	9.740672	14.740672	2.451917
x_8	2.45192	9.807668	14.807668	2.455626
x_9	2.45563	9.822504	14.822504	2.456446
x_{10}	2.45645	9.825783	14.825783	2.456627
x_{11}	2.45663	9.826508	14.826508	2.456667
x_{12}	2.45667	9.826668	14.826668	2.456676
x_{13}	2.45668	9.826703	14.826703	2.456678
x_{14}	2.45668	9.826711	14.826711	2.456678

Again, the system converges at 2.4567 (4dp)

FIRST ORDER PROCESS

If rearranging an equation f(x)=0 to the form x = F(x) leads to x = a as a solution for the iterative sequence $x_{n+1} = F(x_n)$, then f(a)=0.

Each iterate differs from a by an error e_n
so $x_n = a + e_n$

$x_{n+1} = F(x_n)$ becomes $a + e_{n+1} = F(a + e_n)$

Using Taylor's series

$$a + e_{n+1} = F(a) + e_n F'(a) + \frac{e_n^2 F''(a)}{2!} + \frac{e_n^3 F'''(a)}{3!} + ...$$

For convergence,

$$e_{n+1} = e_n F'(a) \quad , F'(a) \neq 0$$

F'(a) is a constant.

This is a first order process.

$$|e_{n+1}| < |e_n|$$
$$\text{converges only if}$$
$$|F'(a)| < 1$$

To find $|F'(a)|$ from consecutive iterates

$$e_n = e_{n-1} F'(a)$$
$$e_{n+1} = e_n F'(a)$$
$$e_{n-1} = \frac{e_n}{F'(a)}$$

$$F'(a) = \frac{x_{n+1} - x_n}{x_n - x_{n+1}}$$
$$= \frac{e_n F'(a) - e_n}{e_n - \frac{e_n}{F'(a)}}$$
$$= \frac{F'(a)(F'(a) - 1)}{(F'(a) - 1)}$$

SECOND ORDER PROCESS

$$\boxed{a + e_{n+1} = F(a) + e_n F'(a) + \frac{e_n^2 F''(a)}{2!} + \frac{e_n^3 F'''(a)}{3!} + \ldots}$$

If F'(a) = 0

$$e_{n+1} = \frac{e_n^2 F''(a)}{2!} \quad , F''(a) \neq 0$$

F"(a) is a constant.

This is a second order process.

RULE OF FALSE POSITION

Used to find approximate roots of the equation f(x) = 0
Take two points on the graph, on either side of the x-axis.

A(a,f(a)) is a fixed point, $P_0(x_0,f(x_0))$ is a varying point.

An iterative equation can then be formed: -

$$x_{n+1} = a - f(a)\frac{a - x_n}{f(a) - f(x_n)}$$

Example

Use the rule of false position to solve the equation
$x^4 - 2x = 0$, x > 0 to two decimal places.

f(x) = x^4 - 2x

f(1) = 1-2=-1 giving (1 , -1)
f(2) = 1-2=-1 giving (2 , 12)

Taking A(2 ,12) and P_0(1 ,-1)

$$x_{n+1} = 2 - 12\left(\frac{2 - x_n}{12 - f(x_n)}\right)$$

$$= 2 - 12\left(\frac{2 - x_n}{12 - x^4 + 2x}\right)$$

Starting with $x_0 = 1$ and $f(x_0) = -1$

$$x_1 = 2 - 12\left(\frac{2-1}{12-(-1)}\right)$$

$$= 2 - \frac{12}{13}$$

$$= 1.0769231$$

$$x_2 = 2 - 12\left(\frac{2-x_1}{12-x^4+2x}\right)$$

$$= 2 - 12\left(\frac{2-1.0769231}{12-(1.0769231)^4+2(1.0769231)}\right)$$

$$= 1.13521$$

etc.

	x_n	$2-x_n$	$f(x_n)$			x_{n+1}
x_0	1	1	-1	13	0.923077	1.076923
x_1	1.07692	0.9230769	0.8088	12.8088	0.86479	1.13521
x_2	1.13521	0.8647904	0.60967	12.60967	0.822978	1.177022
x_3	1.17702	0.8229783	0.43477	12.43477	0.794204	1.205796
x_4	1.2058	0.794204	0.29764	12.29764	0.774982	1.225018
x_5	1.22502	0.7749819	0.19803	12.19803	0.762401	1.237599
x_6	1.2376	0.7624005	0.12924	12.12924	0.754277	1.245723
x_7	1.24572	0.754277	0.08328	12.08328	0.749078	1.250922
x_8	1.25092	0.7490782	0.05323	12.05323	0.74577	1.25423
x_9	1.25423	0.7457702	0.03384	12.03384	0.743673	1.256327
x_{10}	1.25633	0.7436731	0.02144	12.02144	0.742347	1.257653
x_{11}	1.25765	0.7423466	0.01356	12.01356	0.741509	1.258491
x_{12}	1.25849	0.7415089	0.00856	12.00856	0.74098	1.25902
x_{13}	1.25902	0.7409803	-0.0054	12.0054	0.740647	1.259353
x_{14}	1.25935	0.740647	0.00341	12.00341	0.740437	1.259563

$x = 1.26$ 2dp

FUNCTIONS

Apps 1.4: Applying algebraic and calculus skills to properties of functions.

MODULUS FUNCTION

The modulus function $y = |x|$

$|x| = \begin{cases} x & \text{when } x \geq 0 \\ -x & \text{when } x < 0 \end{cases}$ This always gives a positive result.

Example $y = 3x^2 + 6x - 2$ has graph

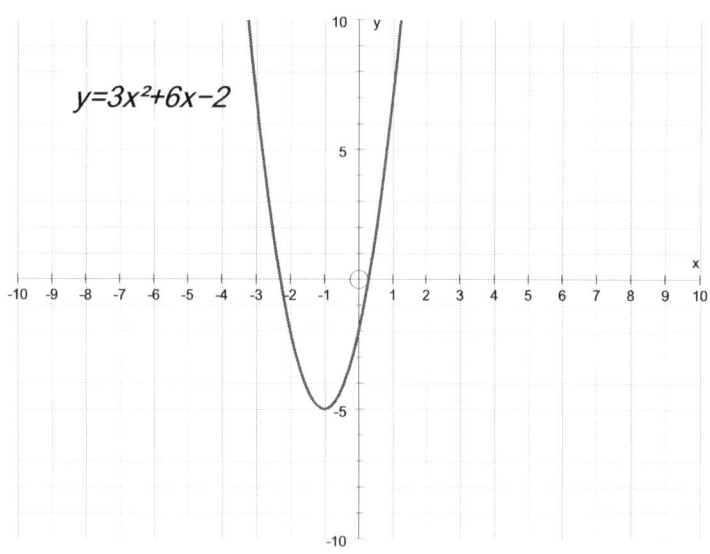

Whereas $y = |3x^2 + 6x - 2|$ has graph

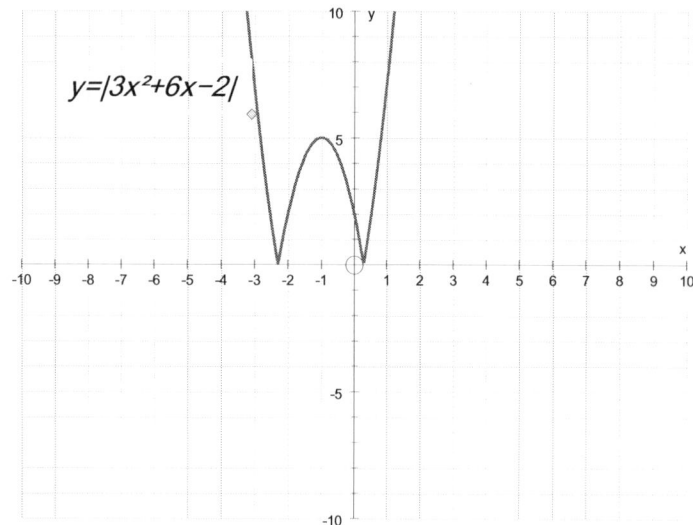

Note how the negative portions have been reflected in the x-axis.

ODD AND EVEN FUNCTIONS

Odd functions have half-turn symmetry about the origin, so $f(-x) = -f(x)$

Examples

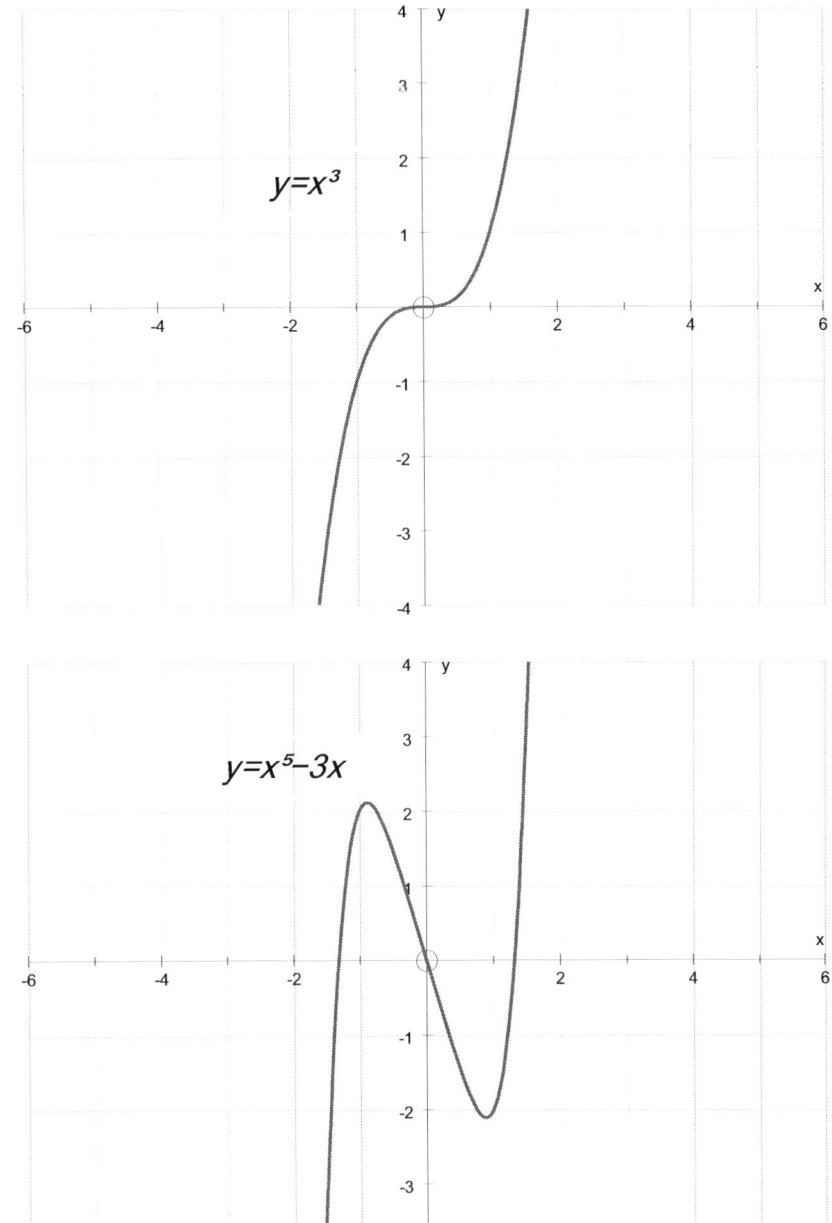

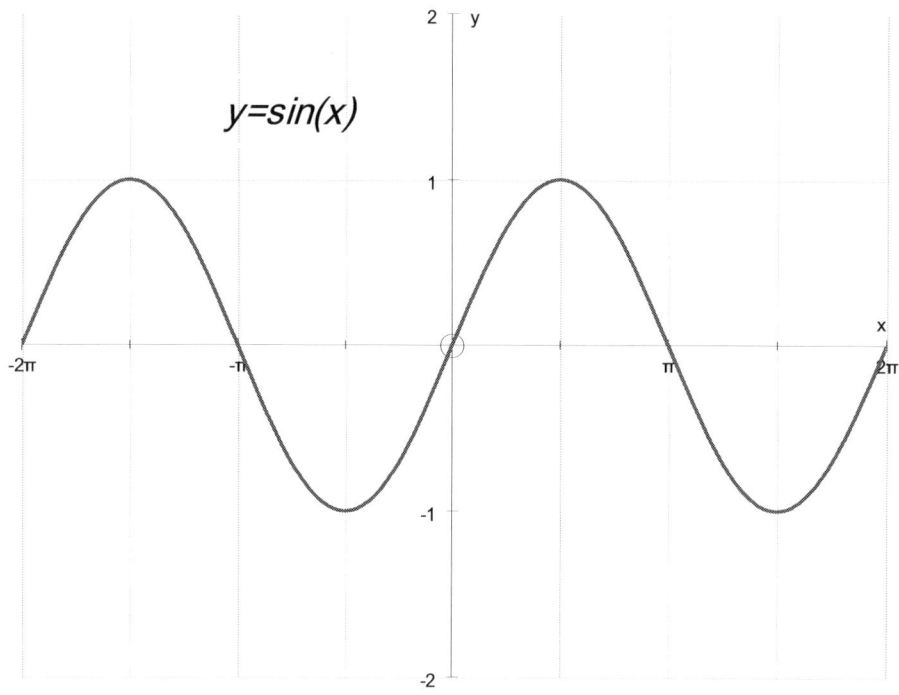

Example

Show that x⁵ + 3x³ is an odd function.

$$f(x) = x^5 + 3x^3$$
$$f(-x) = (-x)^5 + 3(-x)^3$$
$$= -x^5 - 3x^3$$
$$= -(x^5 + 3x^3)$$
$$= -f(x)$$

$f(-x) = -f(x)$ so $x^5 + 3x^3$ is an odd function

Even functions are symmetrical about the y – axis   so   f(-x) = f(x)

Examples

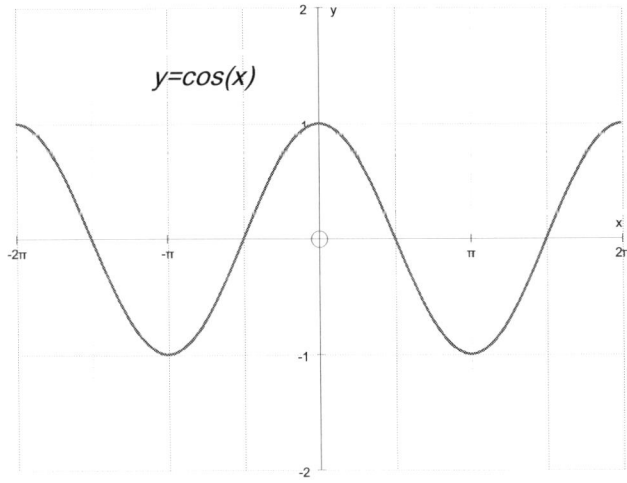

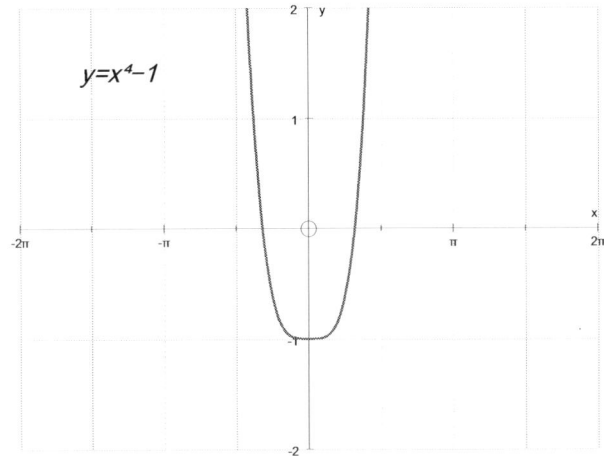

Is $x^6 + 3x^2$ an even function ?

$$f(x) = x^6 + 3x^2$$
$$f(-x) = (-x)^6 + 3(-x)^2$$
$$= x^6 + 3x^2$$
$$= f(x)$$

$f(-x) = f(x)$ so $x^6 + 3x^2$ is an even function

ASYMPTOTES

An asymptote to a curve is a straight line that the curve approaches but never reaches.

Example $f(x) = (x+1)(x-3)/(x+3)(x-4)$

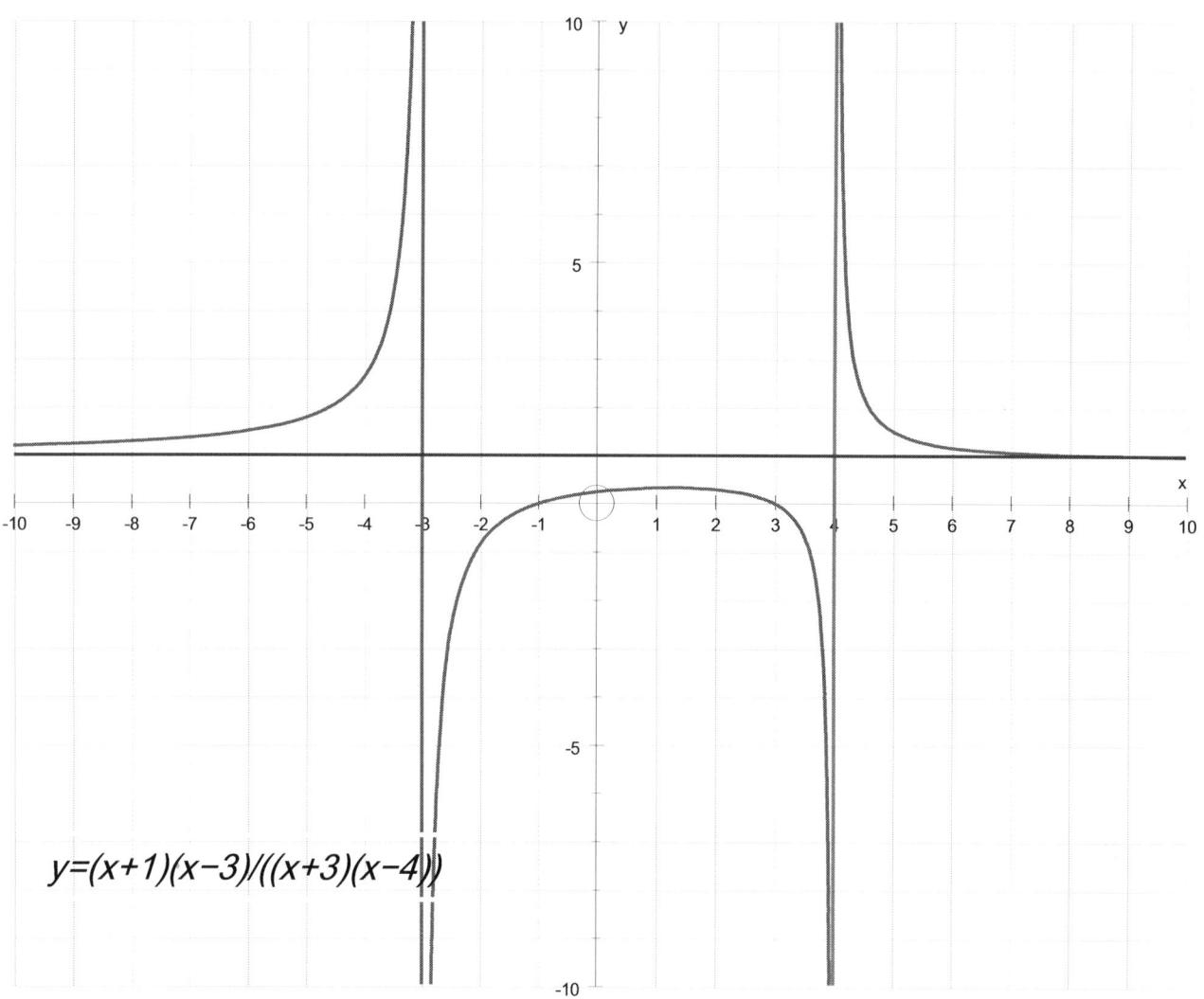

The graph has vertical asymptotes $x = -3$ and $x = 4$
and horizontal asymptote $y = 1$

FINDING ASYMPTOTES

Vertical asymptotes are found by considering what makes the denominator zero. Horizontal and oblique asymptotes need a little further action.

If $f(x) = g(x) + \dfrac{m(x)}{n(x)}$, where $\dfrac{m(x)}{n(x)}$ is a proper rational function

then as $x \to \infty$, $n(x) \to \infty$

$$f(x) \to g(x) \quad \text{and} \quad \dfrac{m(x)}{n(x)} \to 0$$

$g(x)$ is the asymptote of $f(x)$

The curve y=f(x) cuts the asymptote when $\dfrac{m(x)}{n(x)} = 0$

Use algebraic division to reduce the function. The quotient becomes the asymptote.

Example
Find the asymptotes of the function

$$f(x) = \dfrac{(x+2)(x-3)}{(x+1)}$$

$$f(x) = \dfrac{(x+2)(x-3)}{(x+1)}$$

has vertical asymptote x=-1

To find other asymptotes

$$f(x) = \dfrac{(x+2)(x-3)}{(x+1)}$$

$$f(x) = \dfrac{x^2 - x - 6}{(x+1)}$$

$$f(x) = x - 2 - \dfrac{4}{(x+1)}$$

The function has oblique asymptote y = x-2

Alternatively: -
- Asymptotes parallel to the x-axis can be found by equating the co-efficient of the highest power of x to zero.
- Those parallel to the y-axis can be found by equating the co-efficient of the highest power of y to zero.
- To find oblique asymptotes, substitute y = mx + c into the equation and equate the co-efficients of the two highest powers of x to zero.

Example

$$f(x) = \frac{(x+2)(x-3)}{(x+1)}$$

$$y = \frac{(x+2)(x-3)}{(x+1)}$$

$y(x+1) = x^2 - x - 6$

$y(x+1) - x^2 + x + 6 = 0$

Equate coefficients of y to 0

x+1=0, $\Rightarrow$ x=-1

The function has vertical asymptote x= -1

$y(x+1) = x^2 - x - 6$

$yx + y - x^2 + x + 6 = 0$ substitute y=mx+c

$(mx+c)x + mx+c - x^2 + x + 6 = 0$

$mx^2 + cx + mx - x^2 + x + 6 = 0$

$(m-1)x^2 + (c+m+1)x + 6 = 0$

Equate coefficients of x^2 and x to 0

m-1=0   c+m+1=0

$\Rightarrow$ m=1 $\Rightarrow$ c=-2

The function has oblique asymptote y = x-2

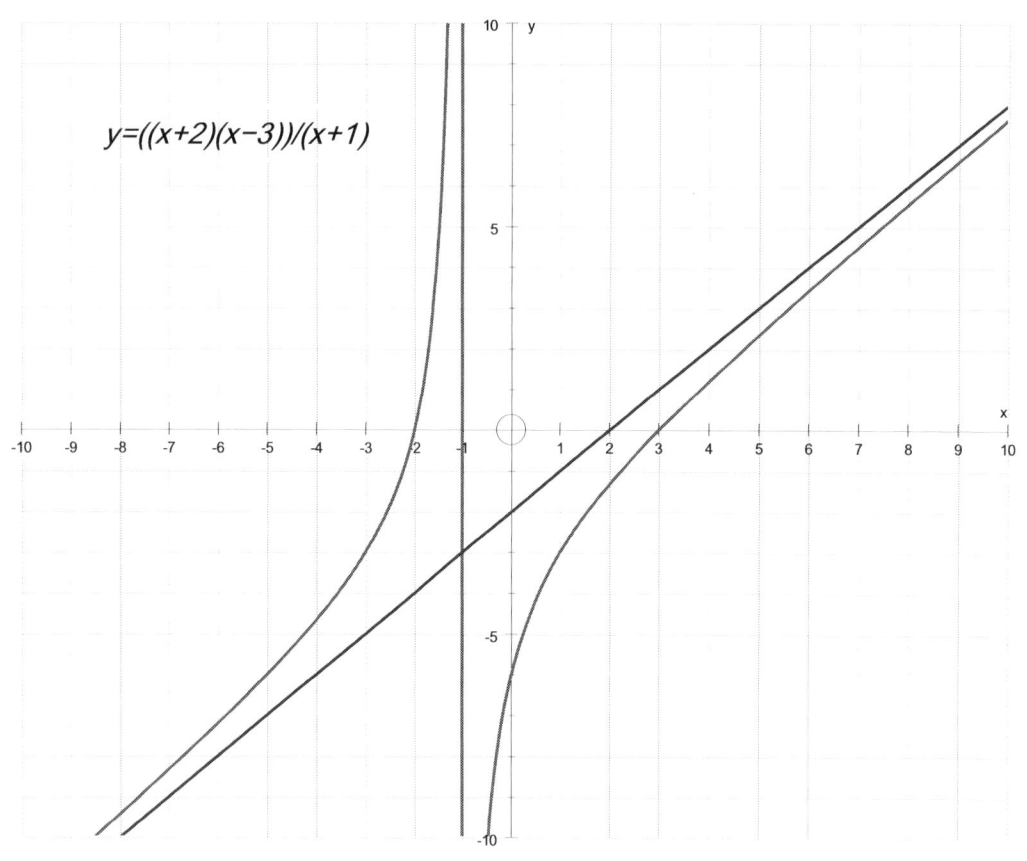

y=((x+2)(x-3))/(x+1)

CRITICAL POINTS

> Critical points of a function occur when
> $$f'(a) = 0$$
> or when $f'(a)$ does not exist

If $(a, f(a))$ is a critical point,

a is called a critical number of the function,

$f(a)$ is called a critical value of the function.

A local maximum value $f(a)$ exists at a if, and only if, an interval centered at a in the domain exists such that $f(a) \geq f(x)$ for all x in the interval.

A local minimum value $f(a)$ exists at a if, and only if, an interval centered at a in the domain exists such that $f(a) \leq f(x)$ for all x in the interval.

An endpoint maximum value $f(a)$ exists at a if $(a, f(a))$ is an endpoint and an interval centered at a in the domain exists such that $f(a) \geq f(x)$ for all x in the interval.

An endpoint minimum value $f(a)$ exists at a if $(a, f(a))$ is an endpoint and an interval centered at a in the domain exists such that $f(a) \leq f(x)$ for all x in the interval.

A global maximum value $f(a)$ exists at a if, and only if,
an interval centered at a in the domain exists such that
$f(a) \geq f(x)$ for all x **in the domain**.

i.e if it is the largest value for the domain being considered.

A global minimum value $f(a)$ exists at a if, and only if,
an interval centered at a in the domain exists such that
$f(a) \leq f(x)$ for all x in the domain.

i.e if it is the smallest value for the domain being considered.

A continuous function defined in a closed interval must have both a global maximum and minimum.

They are found by examining the local extrema and end points.

Find the critical points
List the endpoints under consideration
Global maxima and minima must be one of these.

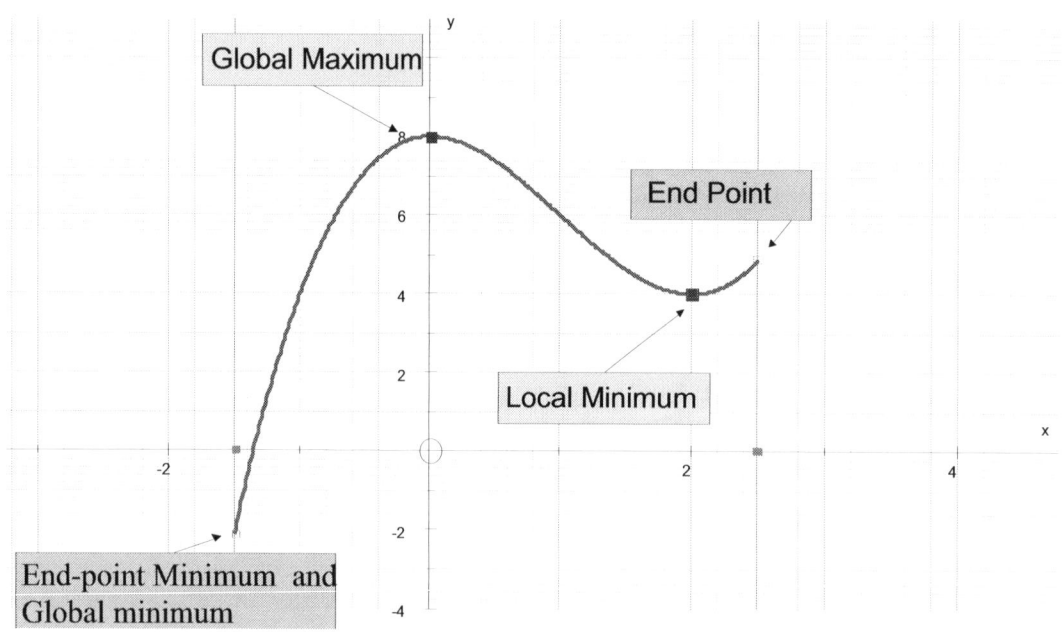

Example

Find the global extrema of the following piecewise function in the domain [-3,2].

$$f(x) = \begin{cases} x+5 & \text{when } -3 \leq x < -2 \\ x^2-1 & \text{when } -2 \leq x < -1 \\ 1-x^2 & \text{when } -1 \leq x < 1 \\ x^2-1 & \text{when } 1 \leq x < 2 \end{cases}$$

so

$$f'(x) = \begin{cases} 1 & \text{when } -3 \leq x < -2 \\ 2x & \text{when } -2 \leq x < -1 \\ -2x & \text{when } -1 \leq x < 1 \\ 2x & \text{when } 1 \leq x < 2 \end{cases}$$

Critical points occur when $f'(a) = 0$
or $f'(a)$ does not exist for any given point $(a, f(a))$

$x = 0$ is a stationary point.
$f'(x) = -2x$,
$f''(0) = -2$, so $(0,1)$ is a local maximum

The critical points to consider are -3,-2,-1,1 and 2dy

$x = -3$ is an endpoint, with no left derivative.
$f(-3) = 2 \quad f(-2) = 3$
$(-3, 2)$ is an endpoint minimum, since it is an endpoint and has the smallest value for the interval $-3 \leq x < 2$

$x = -2$ has a left derivative of 1 and a right derivative -4
$f(-2) = 3$ and $f(-1) = 2$ so $(-2, 3)$ is a local maximum

$x = -1$ has a left derivative of -2 and a right derivative 2
$f(-1) = 0$ and $f(0) = 1$ so $(-1, 0)$ is a local minimum

$x = 1$ has a left derivative of -2 and a right derivative 2
$f(1) = 0$ and $f(1.99) = 2.9601$ so $(1, 0)$ is a local minimum

$x = 2$ has no right derivative.
    It is not an endpoint since it is excluded from the interval.

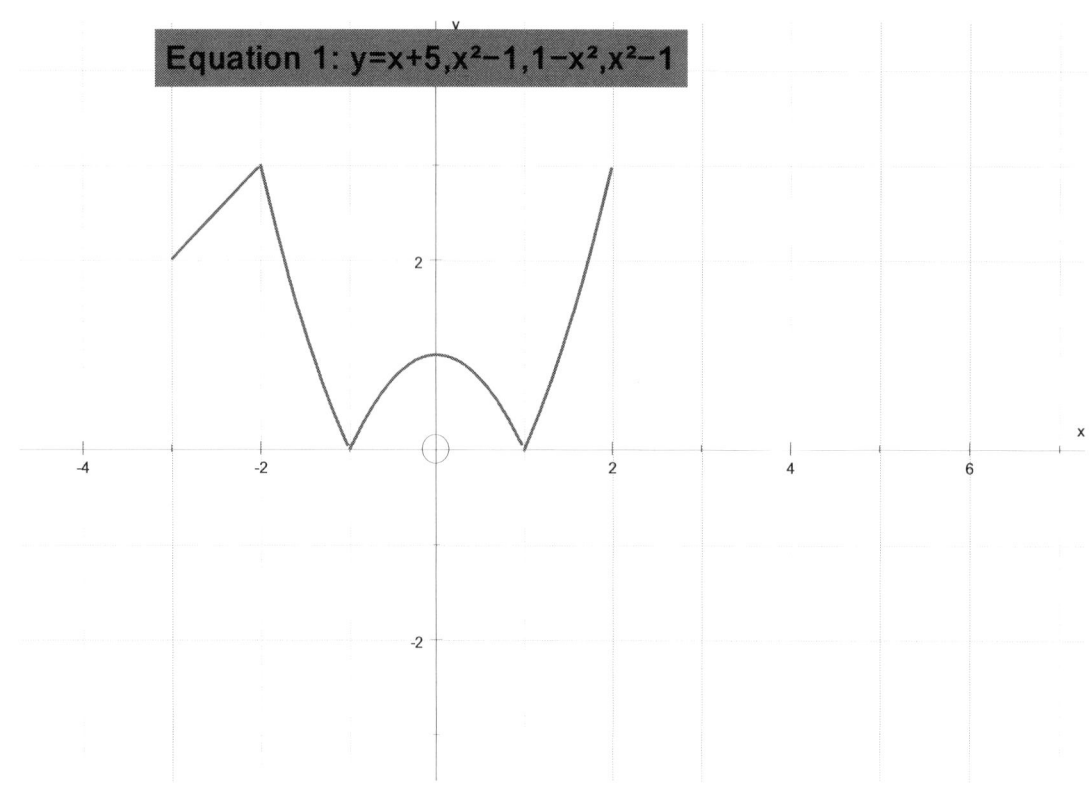

MATRICES

GPS 1.1: Applying algebraic skills to matrices and systems of equations

TYPES

A matrix is a rectangular array of elements (real or complex numbers) arranged in rows and columns. The array is enclosed within box brackets or parentheses.

$$\begin{bmatrix} 2 & 1 \\ 4 & 5 \\ 8 & 9 \\ 1 & 2 \end{bmatrix} \qquad \begin{pmatrix} 2 & 1 \\ 4 & 5 \\ 8 & 9 \\ 1 & 2 \end{pmatrix}$$

box brackets   parentheses

A matrix with m rows and n columns is called an m by n or m x n matrix. This is also called its order.

If m = n, the matrix is called a square matrix.
If m = 1, the matrix is called a row matrix, or row vector.
If n = 1, the matrix is called a column matrix, or column vector.

Examples

$$(1 \quad 5 \quad 8) \qquad \begin{pmatrix} 1 \\ 2 \\ 4 \end{pmatrix}$$

1 x 3 row matrix   3 x 1 column matrix
or row vector   or column vector

$$\begin{pmatrix} 2 & 3 \\ 2 & 31 \\ 15 & 8 \\ 1 & \sqrt{48} \end{pmatrix} \qquad \begin{pmatrix} 1 & 2 & 3 \\ 4 & 5 & 6 \\ 0.8 & 12 & 3i \end{pmatrix}$$

4 x 2 matrix   3 x 3 square matrix

ELEMENTS

Each element in a matrix is addressed by its row and column as double suffixes.

a_{ij} denotes the element in the ith row and j^{th} column.

The elements of an m x n matrix:

$$\begin{pmatrix} a_{11} & a_{12} & a_{13} & \cdots & a_{1n} \\ a_{21} & a_{22} & a_{23} & \cdots & a_{2n} \\ a_{31} & a_{32} & a_{33} & \cdots & a_{3n} \\ \vdots & \vdots & \vdots & & \vdots \\ a_{m1} & a_{m2} & a_{m3} & & a_{mn} \end{pmatrix}$$

a_{23} = the element in the second row and third column

a_{mn} = the element in the mth row and n^{th} column

Example

$$\begin{pmatrix} 1 & 2 & 3 \\ 4 & 5 & 6 \\ 0.8 & 12 & 3i \end{pmatrix}$$

$a_{13}=3$ $a_{21}=4$ a_{34} does not exist

MATRIX NOTATION

An m x n matrix is often denoted by a capital letter in bold.
In written work, a wavy line is drawn under the capital letter.
It can also be denoted by a general element enclosed in brackets.

$$\begin{pmatrix} a_{11} & a_{12} & a_{13} \\ a_{21} & a_{22} & a_{23} \\ a_{31} & a_{32} & a_{33} \end{pmatrix}$$

Can be denoted as **A**, $\underset{\sim}{A}$, (aij) or (a)

A row or column matrix is often denoted by a lowercase letter in bold.

$(y_1 \quad y_2 \quad y_3)$ can be denoted as **y**, $\underset{\sim}{y}$, (yj) or (y)

Equality

Two matrices are equal only if they are of the same order and contain corresponding elements.

$$\begin{bmatrix} 1 & 2 & 3 \\ 4 & 5 & 6 \\ 0.8 & 12 & 3i \end{bmatrix} = \begin{bmatrix} 1 & 2 & 3 \\ 4 & 5 & 6 \\ 0.8 & 12 & 3i \end{bmatrix}$$

$$\begin{bmatrix} 1 & 2 & 3 \\ 4 & 5 & 6 \\ 0.8 & 12 & 3i \end{bmatrix} \neq \begin{bmatrix} 1 & 4 & 0.8 \\ 2 & 5 & 12 \\ 3 & 6 & 3i \end{bmatrix}$$

MATRIX ADDITION & SUBTRACTION

Only matrices of the same order can be added or subtracted.
Each element is added or subtracted to its corresponding element.

Matrix addition is commutative

$$\boxed{A + B = B + A}$$

Examples

$$\begin{bmatrix} 1 & 2 & 3 \\ 4 & 5 & 6 \\ 0.8 & 12 & 3 \end{bmatrix} + \begin{bmatrix} 7 & 21 & 3 \\ 5 & 2 & 1 \\ 0 & 1 & 4 \end{bmatrix} = \begin{bmatrix} 1+7 & 2+21 & 3+3 \\ 4+5 & 5+2 & 6+1 \\ 0.8+0 & 12+1 & 3+4 \end{bmatrix}$$

$$= \begin{bmatrix} 8 & 23 & 6 \\ 9 & 7 & 7 \\ 0.8 & 13 & 7 \end{bmatrix}$$

$$\begin{bmatrix} 7 & 1 \\ 2 & 3 \end{bmatrix} - \begin{bmatrix} 4 & 2 \\ 3 & 1 \end{bmatrix} = \begin{bmatrix} 7-4 & 1-2 \\ 2-3 & 3-1 \end{bmatrix}$$

$$= \begin{bmatrix} 3 & -1 \\ -1 & 2 \end{bmatrix}$$

$$\begin{bmatrix} 5 & 6 \\ 4 & 5 \\ 7 & 8 \end{bmatrix} - \begin{bmatrix} 1 & 2 & 3 \\ 4 & 5 & 6 \end{bmatrix} = \text{Meaningless}$$

SCALAR MATRIX MULTIPLICATION

To multiply a matrix by a scalar, multiply each element by the scalar.

Example

$$7 \times \begin{bmatrix} 7 & 1 \\ 2 & 3 \end{bmatrix} = \begin{bmatrix} 7 \times 7 & 1 \times 7 \\ 2 \times 7 & 3 \times 7 \end{bmatrix} = \begin{bmatrix} 49 & 7 \\ 14 & 21 \end{bmatrix}$$

MATRIX MULTIPLICATION

This can only be done if the number of columns in the first matrix is equal to the number of rows in the second.

i.e. If **A** is an m x n matrix and **B** is as n x p matrix
then the product **A.B** is possible but **B.A** is not possible.

Let **B** be an m x n matrix and **A** be an n x p matrix.

$$\mathbf{B} = \begin{bmatrix} b_{11} & \cdots & b_{1n} \\ \vdots & & \vdots \\ b_{m1} & \cdots & b_{mn} \end{bmatrix} \qquad \mathbf{A} = \begin{bmatrix} a_{11} & \cdots & a_{1p} \\ \vdots & & \vdots \\ a_{n1} & \cdots & a_{np} \end{bmatrix}$$

B.A is possible, **A.B** is not possible

Let the product **BA=C**

$$\mathbf{BA} = \mathbf{C} = \begin{bmatrix} c_{11} & \cdots & c_{1p} \\ \vdots & & \vdots \\ c_{m1} & \cdots & c_{mp} \end{bmatrix}$$

C is defined

$$c_{ij} = \sum_{k=1}^{n} b_{ik} a_{kj} \quad i = 1, 2, \ldots m \quad j = 1, 2, \ldots p$$

$$c_{ij} = b_{i1} a_{1j} + b_{i2} a_{2j} + b_{i3} a_{3j} + \ldots + b_{in} a_{nj}$$

Matrix multiplication could be described as finding the scalar product of each row in the first matrix by each column in the second.

Examples

Let

$$\mathbf{B} = \begin{bmatrix} 3 & 2 & 4 \\ 5 & 6 & 2 \\ 7 & 2 & 7 \end{bmatrix} \qquad \mathbf{A} = \begin{bmatrix} 5 & 1 \\ 2 & 6 \\ 3 & 4 \end{bmatrix}$$

B is a 3x3 matrix   **A is a 3x2 matrix   so B.A** is possible.
A.B is not possible, so multiplication here is not commutative.

$$\boxed{\mathbf{AB} \neq \mathbf{BA}}$$

$$\mathbf{BA} = \mathbf{C} = \begin{bmatrix} 3\times5+2\times2+4\times3 & 3\times1+2\times6+4\times4 \\ 5\times5+6\times2+2\times3 & 5\times1+6\times6+2\times4 \\ 7\times5+2\times2+7\times3 & 7\times1+2\times6+7\times4 \end{bmatrix}$$

$$= \begin{bmatrix} 31 & 31 \\ 43 & 49 \\ 60 & 47 \end{bmatrix}$$

Note

The product of an m x n matrix with an n x p matrix is a matrix with order m x p

In the product **B.A**
A is pre-multiplied by **B**
B is post-multiplied by **A**

Matrix multiplication is associative

$$\boxed{(\mathbf{AB})\mathbf{C} = \mathbf{A}(\mathbf{BC})}$$

and distributive

$$\boxed{\mathbf{A}(\mathbf{B}+\mathbf{C}) = \mathbf{AB} + \mathbf{AC}}$$

provided B and C have the same order and that the number of columns in A equals the number of rows in B and C.

TRANSPOSE

The transpose of a matrix **A** is written **A'** or **AT**.

It is found by interchanging the rows and columns, so that
$a'_{ij} = a_{ji}$

Example

$$D = \begin{bmatrix} 1 & 2 \\ 3 & 4 \end{bmatrix} \qquad D^T = \begin{bmatrix} 1 & 3 \\ 2 & 4 \end{bmatrix}$$

$$A = \begin{bmatrix} 1 & 2 & 3 \\ 4 & 5 & 6 \\ 0.8 & 12 & 3i \end{bmatrix} \qquad A^T = \begin{bmatrix} 1 & 4 & 0.8 \\ 2 & 5 & 12 \\ 3 & 6 & 3i \end{bmatrix}$$

Notice that the leading diagonal remains the same

$$A = \begin{bmatrix} 1 & 2 & 3 \\ 4 & 5 & 6 \\ 0.8 & 12 & 3i \end{bmatrix} \qquad A^T = \begin{bmatrix} 1 & 4 & 0.8 \\ 2 & 5 & 12 \\ 3 & 6 & 3i \end{bmatrix}$$

and that the other entries have been flipped.

$$\boxed{\left(A^T\right)^T = A}$$

$$\boxed{\left(A+B\right)^T = A^T + B^T}$$

$$\boxed{\left(AB\right)^T = B^T A^T}$$

SPECIAL MATRICES

A square matrix is symmetric if $a_{ij} = a_{ij}$

$$A = \begin{bmatrix} 1 & 2 & 3 \\ 2 & 5 & 6 \\ 3 & 6 & 3i \end{bmatrix} \quad A^T = \begin{bmatrix} 1 & 2 & 3 \\ 2 & 5 & 6 \\ 3 & 6 & 3i \end{bmatrix}$$

A symmetric matrix has transpose $A^T = A$

A square matrix is skew-symmetric if $AT = -A$

$$A = \begin{bmatrix} 0 & -2 & 3 \\ 2 & 0 & 6 \\ -3 & -6 & 0 \end{bmatrix} \quad A^T = \begin{bmatrix} 0 & 2 & -3 \\ -2 & 0 & -6 \\ 3 & 6 & 0 \end{bmatrix}$$

Notice that the leading diagonal is zero

A Diagonal matrix is a square matrix whose non leading diagonal elements are zeroes.

$$A = \begin{bmatrix} 1 & 0 & 0 \\ 0 & 2 & 0 \\ 0 & 0 & 3 \end{bmatrix}$$

The Unit or Identity matrix is a diagonal matrix whose leading diagonal elements are all ones. It is called **I**.

$$I = \begin{bmatrix} 1 & 0 & 0 \\ 0 & 1 & 0 \\ 0 & 0 & 1 \end{bmatrix} \quad \boxed{AI = A}$$

The Null or Zero matrix is a matrix made purely from zeroes.
It is called **0**.

$$0 = \begin{bmatrix} 0 & 0 & 0 \\ 0 & 0 & 0 \\ 0 & 0 & 0 \end{bmatrix}$$

Unlike in ordinary algebra,
A.B = 0 does not allow the assumption that either **A** or **B** is zero.

$$\begin{bmatrix} 1 & 1 & 2 \\ 2 & 2 & 4 \\ 4 & 4 & 8 \end{bmatrix} \begin{bmatrix} 1 & 2 & 4 \\ 3 & 6 & 12 \\ -2 & -4 & -8 \end{bmatrix} = \begin{bmatrix} 0 & 0 & 0 \\ 0 & 0 & 0 \\ 0 & 0 & 0 \end{bmatrix} = 0$$

TRANSFORMATION MATRICES

Linear transformations can be described by a mapping such that T maps the points (x,y) to (x',y')

$$T: \begin{pmatrix} x \\ y \end{pmatrix} \rightarrow \begin{pmatrix} x' \\ y' \end{pmatrix} \quad \text{or} \quad \begin{pmatrix} x' \\ y' \end{pmatrix} = A \begin{pmatrix} x \\ y \end{pmatrix}$$

where A (an mxn matrix) is called the transformation matrix of T.

Example

T:x' = ax+by, y'=cx+dy can be written

$$\begin{pmatrix} x' \\ y' \end{pmatrix} = \begin{pmatrix} ax + by \\ cx + dy \end{pmatrix}$$

$$= \begin{pmatrix} a & b \\ c & d \end{pmatrix} \begin{pmatrix} x \\ y \end{pmatrix}$$

$$= A \begin{pmatrix} x \\ y \end{pmatrix}$$

where $A = \begin{pmatrix} a & b \\ c & d \end{pmatrix}$

Example

Find the images of the points A(1,2), B(3,4) and C(5,6) after it is translated by the matrix $\begin{pmatrix} 1 & 2 \\ 3 & 4 \end{pmatrix}$

$$\begin{pmatrix} x' \\ y' \end{pmatrix} = \begin{pmatrix} 1 & 2 \\ 3 & 4 \end{pmatrix} \begin{pmatrix} 1 & 3 & 5 \\ 2 & 4 & 6 \end{pmatrix}$$

$$= \begin{pmatrix} 5 & 11 & 17 \\ 11 & 25 & 39 \end{pmatrix}$$

A'(5,11)   B'(11,25)   C'(17,39)

REFLECTION AND ROTATION

Reflection in x axis

$$A = \begin{pmatrix} 1 & 0 \\ 0 & -1 \end{pmatrix}$$

Reflection in y axis

$$A = \begin{pmatrix} -1 & 0 \\ 0 & 1 \end{pmatrix}$$

Reflection in line y=x

$$A = \begin{pmatrix} 0 & 1 \\ 1 & 0 \end{pmatrix}$$

Reflection in line y=-x

$$A = \begin{pmatrix} 0 & -1 \\ -1 & 0 \end{pmatrix}$$

Rotation 90° anticlockwise about the origin.

$$A = \begin{pmatrix} 0 & -1 \\ 1 & 0 \end{pmatrix}$$

Rotation 180° about the origin.

$$A = \begin{pmatrix} -1 & 0 \\ 0 & -1 \end{pmatrix}$$

Rotation 90° clockwise about the origin.

$$A = \begin{pmatrix} 0 & -1 \\ 1 & 0 \end{pmatrix}$$

Rotation θ anticlockwise about the origin.

$$A = \begin{pmatrix} \cos\theta & -\sin\theta \\ \sin\theta & \cos\theta \end{pmatrix}$$

x' = xcosθ − ysinθ   y'= xsinθ + ycosθ

ENLARGEMENT WITH SCALE FACTOR

Examples

$$A = \begin{pmatrix} 2 & 0 \\ 0 & 2 \end{pmatrix} \qquad A = \begin{pmatrix} 1/2 & 0 \\ 0 & 1/2 \end{pmatrix}$$

Scaling

$$A = \begin{pmatrix} \lambda & 0 \\ 0 & \mu \end{pmatrix}$$

λ = scaling in x direction
μ = scaling in y direction

To undo the effects of a transformation,
premultiply the image vector by the inverse of the transformation matrix.

If the inverse of the transformation matrix equals its transpose, then the transformation matrix is orthogonal.

If a point is transformed to its own image, it is called invariant.

DETERMINANTS

The system of equations
ax +by =e
cx + dy =f

can be written in matrix form as **Ax=b**

Where $\mathbf{A} = \begin{bmatrix} a & b \\ c & d \end{bmatrix}$ $\mathbf{x} = \begin{bmatrix} x \\ y \end{bmatrix}$ and $\mathbf{b} = \begin{bmatrix} e \\ f \end{bmatrix}$

Notice that the denominator from above, $ad - bc$, is found by cross multiplying and subtracting the elements in **A**.

This is known as the determinant of A.

It is written det(A) or $|\mathbf{A}|$

$\mathbf{A} = \begin{pmatrix} a & b \\ c & d \end{pmatrix}$ has determinant $\begin{vmatrix} a & b \\ c & d \end{vmatrix} = ad - bc$

If det(A) = 0, the system has no solution.

Example

Does the following system of equations have a solution?

3x +2y = 7
9x + 6y = 6

$\begin{vmatrix} 3 & 2 \\ 9 & 6 \end{vmatrix} = 3 \times 6 - 9 \times 2 = 0$

No solution, since determinant = 0

DETERMINANT OF A 3X3 MATRIX

$$A = \begin{pmatrix} a_1 & b_1 & c_1 \\ a_2 & b_2 & c_2 \\ a_3 & b_3 & c_3 \end{pmatrix}$$

$$\det(A) = \begin{vmatrix} a_1 & b_1 & c_1 \\ a_2 & b_2 & c_2 \\ a_3 & b_3 & c_3 \end{vmatrix}$$

Each element in A is associated with its minor

$$\begin{vmatrix} a_1 & b_1 & c_1 \\ a_2 & b_2 & c_2 \\ a_3 & b_3 & c_3 \end{vmatrix} \quad \text{The minor of } a_1 \text{ is} \begin{vmatrix} b_2 & c_2 \\ b_3 & c_3 \end{vmatrix}$$

$$= b_2 c_3 - b_3 c_2$$

$$\begin{vmatrix} a_1 & b_1 & c_1 \\ a_2 & b_2 & c_2 \\ a_3 & b_3 & c_3 \end{vmatrix} \quad \text{The minor of } b_1 \text{ is} \begin{vmatrix} a_2 & c_2 \\ a_3 & c_3 \end{vmatrix}$$

$$= a_2 c_3 - a_3 c_2$$

$$\begin{vmatrix} a_1 & b_1 & c_1 \\ a_2 & b_2 & c_2 \\ a_3 & b_3 & c_3 \end{vmatrix} \quad \text{The minor of } c_1 \text{ is} \begin{vmatrix} a_2 & b_2 \\ a_3 & b_3 \end{vmatrix}$$

$$= a_2 b_3 - a_3 b_2$$

To find the determinant,

write down each element from the top row.

Multiply it by its minor.

Add or subtract alternatively according to its position

Place signs

$$\begin{vmatrix} + & - & + \\ - & + & - \\ + & - & + \end{vmatrix}$$

$$\det(\mathbf{A}) = \begin{vmatrix} a_1 & b_1 & c_1 \\ a_2 & b_2 & c_2 \\ a_3 & b_3 & c_3 \end{vmatrix}$$

$$= a_1 \begin{vmatrix} b_2 & c_2 \\ b_3 & c_3 \end{vmatrix} - b_1 \begin{vmatrix} a_2 & c_2 \\ a_3 & c_3 \end{vmatrix} + c_1 \begin{vmatrix} a_2 & b_2 \\ a_3 & b_3 \end{vmatrix}$$

$\det(\mathbf{AB}) = \det \mathbf{A} \det \mathbf{B}$

Example

Find the determinant of $A = \begin{pmatrix} 2 & 3 & 5 \\ 4 & 1 & 6 \\ 1 & 3 & 0 \end{pmatrix}$

$$\det(A) = \begin{vmatrix} 2 & 3 & 5 \\ 4 & 1 & 6 \\ 1 & 3 & 0 \end{vmatrix}$$

$$= 2\begin{vmatrix} 1 & 6 \\ 3 & 0 \end{vmatrix} - 3\begin{vmatrix} 4 & 6 \\ 1 & 0 \end{vmatrix} + 5\begin{vmatrix} 4 & 1 \\ 1 & 3 \end{vmatrix}$$

$$= 2(1 \times 0 - 3 \times 6) - 3(4 \times 0 - 1 \times 6) + 5(4 \times 3 - 1 \times 1)$$
$$= 2 \times (-18) - 3 \times (-6) + 5 \times 11$$
$$= -36 + 18 + 55$$
$$= 37$$

COFACTORS

Cofactors are minors with their place sign.

Example

Find the cofactors of $A = \begin{pmatrix} 2 & 3 & 5 \\ 4 & 1 & 6 \\ 1 & 3 & 0 \end{pmatrix}$

$A = \begin{pmatrix} 2 & 3 & 5 \\ 4 & 1 & 6 \\ 1 & 3 & 0 \end{pmatrix}$ $sign \begin{pmatrix} + & - & + \\ - & + & - \\ + & - & + \end{pmatrix}$

$\begin{vmatrix} 2 & 3 & 5 \\ 4 & 1 & 6 \\ 1 & 3 & 0 \end{vmatrix}$ The minor of a_1 is $\begin{vmatrix} 1 & 6 \\ 3 & 0 \end{vmatrix}$

$= (0 - 18) = -18$

The cofactor of a_1 is $+(-18) = -18$

$\begin{vmatrix} 2 & 3 & 5 \\ 4 & 1 & 6 \\ 1 & 3 & 0 \end{vmatrix}$ The minor of a_2 is $\begin{vmatrix} 3 & 5 \\ 3 & 0 \end{vmatrix}$

$= 0 - 15 = -15$

The cofactor of a_2 is $-(-15) = 15$

$\begin{vmatrix} 2 & 3 & 5 \\ 4 & 1 & 6 \\ 1 & 3 & 0 \end{vmatrix}$ The minor of b_1 is $\begin{vmatrix} 4 & 6 \\ 1 & 0 \end{vmatrix}$

$= 0 - 6 = -6$

The cofactor of b_1 is $-(-6) = 6$

$\begin{vmatrix} 2 & 3 & 5 \\ 4 & 1 & 6 \\ 1 & 3 & 0 \end{vmatrix}$ The minor of b_2 is $\begin{vmatrix} 2 & 5 \\ 1 & 0 \end{vmatrix}$

$= 0 - 5 = -5$

The cofactor of b_2 is $+(-5) = -5$

$\begin{vmatrix} 2 & 3 & 5 \\ 4 & 1 & 6 \\ 1 & 3 & 0 \end{vmatrix}$ The minor of c_1 is $\begin{vmatrix} 4 & 1 \\ 1 & 3 \end{vmatrix}$

$= 12 - 1 = 11$

The cofactor of c_1 is $+(11) = 11$

$\begin{vmatrix} 2 & 3 & 5 \\ 4 & 1 & 6 \\ 1 & 3 & 0 \end{vmatrix}$ The minor of c_2 is $\begin{vmatrix} 2 & 3 \\ 1 & 3 \end{vmatrix}$

$= 6 - 3 = 3$

The cofactor of c_1 is $-(3) = -3$

$\begin{vmatrix} 2 & 3 & 5 \\ 4 & 1 & 6 \\ 1 & 3 & 0 \end{vmatrix}$ The minor of a_3 is $\begin{vmatrix} 3 & 5 \\ 1 & 6 \end{vmatrix}$

$= 18 - 5 = 13$

The cofactor of a_3 is $+(13) = 13$

$\begin{vmatrix} 2 & 3 & 5 \\ 4 & 1 & 6 \\ 1 & 3 & 0 \end{vmatrix}$ The minor of b_3 is $\begin{vmatrix} 2 & 5 \\ 4 & 6 \end{vmatrix}$

$= 12 - 20 = -8$

The cofactor of b_3 is $-(-8) = 8$

$$\begin{vmatrix} 2 & 3 & 5 \\ 4 & 1 & 6 \\ 1 & 3 & 0 \end{vmatrix} \quad \text{The minor of } c_3 \text{ is } \begin{vmatrix} 2 & 3 \\ 4 & 1 \end{vmatrix}$$

$$= 2 - 12 = -10$$

The cofactor of c_3 is $+(-10) = -10$

Combining these with the will give a new matrix C, made up of the cofactors of A.

$$A = \begin{pmatrix} 2 & 3 & 5 \\ 4 & 1 & 6 \\ 1 & 3 & 0 \end{pmatrix} \quad C = \begin{pmatrix} -18 & 6 & 11 \\ 15 & -5 & -3 \\ 13 & 8 & -10 \end{pmatrix}$$

ADJOINT

The Adjoint of A is written Adj (A)
It is the transpose of **C**

$$\boxed{Adj(A) = C^T}$$

Example
Find Adj (**A**) for the matrix

$$A = \begin{pmatrix} 2 & 3 & 5 \\ 4 & 1 & 6 \\ 1 & 3 & 0 \end{pmatrix}$$

has cofactor matrix

$$C = \begin{pmatrix} -18 & 6 & 11 \\ 15 & -5 & -3 \\ 13 & 8 & -10 \end{pmatrix}$$

$$C^T = \begin{pmatrix} -18 & 15 & 13 \\ 6 & -5 & 8 \\ 11 & -3 & -10 \end{pmatrix}$$

$$A = \begin{pmatrix} 2 & 3 & 5 \\ 4 & 1 & 6 \\ 1 & 3 & 0 \end{pmatrix}$$

$$\therefore Adj(A) = C^T = \begin{pmatrix} -18 & 15 & 13 \\ 6 & -5 & 8 \\ 11 & -3 & -10 \end{pmatrix}$$

INVERSE OF A SQUARE MATRIX

$$A^{-1} = \frac{adj(A)}{\det(A)}$$

Example

Find the inverse of the matrix

$$A = \begin{pmatrix} 2 & 3 & 5 \\ 4 & 1 & 6 \\ 1 & 3 & 0 \end{pmatrix}$$

From the previous work,

$$Adj(A) = \begin{pmatrix} -18 & 15 & 13 \\ 6 & -5 & 8 \\ 11 & -3 & -10 \end{pmatrix}$$

$\det(A) = 37$

$$\therefore A^{-1} = \frac{adj(A)}{\det(A)}$$

$$= \frac{1}{37} \begin{pmatrix} -18 & 15 & 13 \\ 6 & -5 & 8 \\ 11 & -3 & -10 \end{pmatrix}$$

$$(AB)^{-1} = B^{-1}A^{-1}$$

PRODUCT OF A SQUARE MATRIX AND ITS INVERSE

$$\boxed{A \cdot A^{-1} = A^{-1} \cdot A = I}$$

from above

$$A \cdot A^{-1} = \begin{pmatrix} 2 & 3 & 5 \\ 4 & 1 & 6 \\ 1 & 3 & 0 \end{pmatrix} \cdot \left[\frac{1}{37} \begin{pmatrix} -18 & 15 & 13 \\ 6 & -5 & 8 \\ 11 & -3 & -10 \end{pmatrix} \right]$$

$$= \frac{1}{37} \begin{pmatrix} 2 & 3 & 5 \\ 4 & 1 & 6 \\ 1 & 3 & 0 \end{pmatrix} \cdot \begin{pmatrix} -18 & 15 & 13 \\ 6 & -5 & 8 \\ 11 & -3 & -10 \end{pmatrix}$$

$$= \frac{1}{37} \begin{pmatrix} -36+18+55 & 30-15-15 & 26+24-50 \\ -72+6+66 & 60-5-18 & 52+8-60 \\ -18+18+0 & 15-15+0 & 13+24-0 \end{pmatrix}$$

$$= \frac{1}{37} \begin{pmatrix} 37 & 0 & 0 \\ 0 & 37 & 0 \\ 0 & 0 & 37 \end{pmatrix}$$

$$= \begin{pmatrix} 1 & 0 & 0 \\ 0 & 1 & 0 \\ 0 & 0 & 1 \end{pmatrix}$$

This means that Gaussian elimination can be used to find the inverse.
Write down the matrix to be inverted with the Identity matrix next to it.

Perform ERO's to reduce the left-hand matrix, simultaneously performing the same ERO's on the Identity matrix.

The result is the inverse of the given matrix.

Example – using the previous question

$$A = \begin{pmatrix} 2 & 3 & 5 \\ 4 & 1 & 6 \\ 1 & 3 & 0 \end{pmatrix}$$

$$= \begin{pmatrix} 2 & 3 & 5 \\ 4 & 1 & 6 \\ 1 & 3 & 0 \end{pmatrix} \begin{pmatrix} 1 & 0 & 0 \\ 0 & 1 & 0 \\ 0 & 0 & 1 \end{pmatrix}$$

$R1 \to \dfrac{1}{2} R1$
$R2 \to 2R1 - R2$
$R3 \to R1 - 2R3$

$$\begin{pmatrix} 1 & 3/2 & 5/2 \\ 0 & 5 & 4 \\ 0 & -3 & 5 \end{pmatrix} \begin{pmatrix} 1/2 & 0 & 0 \\ 2 & -1 & 0 \\ 1 & 0 & -2 \end{pmatrix}$$

$R2 \to \dfrac{1}{5} R2$
$R3 \to R3 + \dfrac{3}{5} R2$

$$\begin{pmatrix} 1 & 3/2 & 5/2 \\ 0 & 1 & 4/5 \\ 0 & 0 & 37/5 \end{pmatrix} \begin{pmatrix} 1/2 & 0 & 0 \\ 2/5 & -1/5 & 0 \\ 11/5 & -3/5 & -2 \end{pmatrix}$$

$R3 \to \dfrac{5}{37} R3$

$$\begin{pmatrix} 1 & 3/2 & 5/2 \\ 0 & 1 & 4/5 \\ 0 & 0 & 1 \end{pmatrix} \begin{pmatrix} 1/2 & 0 & 0 \\ 2/5 & -1/5 & 0 \\ 11/37 & -3/37 & -10/37 \end{pmatrix}$$

$$R3 \to \frac{5}{37}R3 \quad \begin{pmatrix} 1 & 3/2 & 5/2 \\ 0 & 1 & 4/5 \\ 0 & 0 & 1 \end{pmatrix} \begin{pmatrix} 1/2 & 0 & 0 \\ 2/5 & -1/5 & 0 \\ 11/37 & -3/37 & -10/37 \end{pmatrix}$$

$$\begin{aligned} R1 &\to R1 - \frac{5}{2}R3 \\ R2 &\to R2 - \frac{4}{5}R3 \end{aligned} \quad \begin{pmatrix} 1 & 3/2 & 0 \\ 0 & 1 & 0 \\ 0 & 0 & 1 \end{pmatrix} \begin{pmatrix} -9/37 & 15/74 & 25/37 \\ 6/37 & -5/37 & 8/37 \\ 11/37 & -3/37 & -10/37 \end{pmatrix}$$

$$R1 \to R1 - \frac{3}{2}R2 \quad \begin{pmatrix} 1 & 0 & 0 \\ 0 & 1 & 0 \\ 0 & 0 & 1 \end{pmatrix} \begin{pmatrix} -18/37 & 15/37 & 13/37 \\ 6/37 & -5/37 & 8/37 \\ 11/37 & -3/37 & -10/37 \end{pmatrix}$$

$$\text{so } A^{-1} = \begin{pmatrix} -18/37 & 15/37 & 13/37 \\ 6/37 & -5/37 & 8/37 \\ 11/37 & -3/37 & -10/37 \end{pmatrix}$$

$$\text{or } A^{-1} = \frac{1}{37}\begin{pmatrix} -18 & 15 & 13 \\ 6 & -5 & 8 \\ 11 & -3 & -10 \end{pmatrix}$$

GAUSSIAN ELIMINATION

Solving simultaneous equations can be done quite easily by elimination. However, they can also be solved by using matrices.

Example
ax + by = e
cx + dy = f

can be represented by   **Ax=b**

Where $\mathbf{A} = \begin{bmatrix} a & b \\ c & d \end{bmatrix}$ $\mathbf{x} = \begin{bmatrix} x \\ y \end{bmatrix}$ and $\mathbf{b} = \begin{bmatrix} e \\ f \end{bmatrix}$

so (using parenthesis now)

Ax = b

$$\begin{pmatrix} a & b \\ c & d \end{pmatrix} \begin{pmatrix} x \\ y \end{pmatrix} = \begin{pmatrix} e \\ f \end{pmatrix}$$

This can be further shortened to the augmented matrix form by combining A and b

$$\begin{pmatrix} a & b & \vdots & e \\ c & d & \vdots & f \end{pmatrix}$$

that then allows Elementary Row Operations to be used, to reduce the equations to upper triangular form.

$$\begin{pmatrix} g & h & \vdots & j \\ 0 & i & \vdots & k \end{pmatrix}$$ or to continue and solve $$\begin{pmatrix} 1 & 0 & \vdots & n \\ 0 & 1 & \vdots & p \end{pmatrix}$$

Reforming the matrix then gives $$\begin{pmatrix} 1 & 0 \\ 0 & 1 \end{pmatrix} \begin{pmatrix} x \\ y \end{pmatrix} = \begin{pmatrix} n \\ p \end{pmatrix}$$

so $x = n \quad y = p$

Elementary Row Operations.

The order of equations can be switched.

An equation can be multiplied by a constant.

Equations can be added or subtracted.

The process of reducing a system of equations to upper triangular form, then back substituting to solve, is called Gaussian elimination.

A redundant row indicates the lack of a unique solution, a general solution exists.

An inconsistent row indicates that no solution exists.

Small changes in the coefficients leading to large changes in the solution indicate an ill-conditioned system.

Example

Solve the system of equations

$2x + y - 2z = 5$

$3x + 2y + 5z = 5$

$4x + 2y - 4z = 10$

$$\begin{pmatrix} 2 & 1 & -2 & : & 5 \\ 3 & 2 & 5 & : & 5 \\ 4 & 2 & -4 & : & 10 \end{pmatrix}$$

$R1 \to R2 - R1$

$R3 \to R3 - 2R1$

$$\begin{pmatrix} 1 & 1 & 7 & : & 0 \\ 3 & 2 & 5 & : & 5 \\ 0 & 0 & 0 & : & 0 \end{pmatrix}$$

row 3 indicates a general solution

$R2 \to R2 - 3R1$

$$\begin{pmatrix} 1 & 1 & 7 & : & 0 \\ 0 & -1 & -16 & : & 5 \\ 0 & 0 & 0 & : & 0 \end{pmatrix}$$

$R1 \to R1 + R2$

$$\begin{pmatrix} 1 & 0 & -9 & : & 5 \\ 0 & -1 & -16 & : & 5 \\ 0 & 0 & 0 & : & 0 \end{pmatrix}$$

$$\begin{pmatrix} 1 & 0 & -9 \\ 0 & -1 & -16 \\ 0 & 0 & 0 \end{pmatrix} \bullet \begin{pmatrix} x \\ y \\ z \end{pmatrix} = \begin{pmatrix} 5 \\ 5 \\ 0 \end{pmatrix}$$

$x - 9z = 5$

$\therefore x = 5 + 9z$

$-y - 16z = 5$

$z = \dfrac{-y - 5}{16}$

$\therefore x = 5 + 9\left(\dfrac{-y - 5}{16}\right)$

$\Rightarrow x = 5 + \dfrac{-9y - 45}{16}$

$\Rightarrow x = \dfrac{-9y + 35}{16}$

VECTORS

GPS 1.2: Applying algebraic and geometric skills to vectors.

DIRECTION RATIOS AND COSINES

The point $A(a_1, a_2, a_3)$ can be written as postition vector

$$\mathbf{a} = \begin{pmatrix} a_1 \\ a_2 \\ a_3 \end{pmatrix}$$

The ratio $a_1 : a_2 : a_3$ is called the direction ratio.
Parallel vectors have equal direction ratios.

The direction cosines are

$$\mathbf{u} = \begin{pmatrix} \cos\alpha \\ \cos\beta \\ \cos\gamma \end{pmatrix}$$

where $\mathbf{u}$ is a unit vector in the direction of $\mathbf{a}$

with length $|\mathbf{u}_a| = \left|\dfrac{1}{|\mathbf{a}|}\mathbf{a}\right| = \left|\dfrac{1}{|\mathbf{a}|}\right| |\mathbf{a}| = 1$

Example

Show that the points A(-1,-8,-2), B(2, -5, 4) and C(3, -4,6) are collinear, and establish the direction cosines of AB

$$\overrightarrow{AB} = \begin{pmatrix} 2 \\ -5 \\ 4 \end{pmatrix} - \begin{pmatrix} -1 \\ -8 \\ -2 \end{pmatrix} = \begin{pmatrix} 3 \\ 3 \\ 6 \end{pmatrix}$$

$$\overrightarrow{BC} = \begin{pmatrix} 3 \\ -4 \\ 6 \end{pmatrix} - \begin{pmatrix} 2 \\ -5 \\ 4 \end{pmatrix} = \begin{pmatrix} 1 \\ 1 \\ 2 \end{pmatrix}$$

$\overrightarrow{AB}$ has direction ratio $3:3:6 = 1:1:2$

$\overrightarrow{BC}$ has direction ratio $1:1:2$

The direction ratios are the same, so the vectors are parallel

B is common to both AB and BC, so A, B and C are collinear.

$$|\overrightarrow{AB}| = \sqrt{3^2 + 3^2 + 6^2} = \sqrt{54} = 3\sqrt{6}$$

So $\mathbf{u}_{\overrightarrow{AB}} = \dfrac{3}{3\sqrt{6}}\mathbf{i} + \dfrac{3}{3\sqrt{6}}\mathbf{j} + \dfrac{6}{3\sqrt{6}}\mathbf{k}$

$$= \dfrac{1}{\sqrt{6}}\mathbf{i} + \dfrac{1}{\sqrt{6}}\mathbf{j} + \dfrac{2}{\sqrt{6}}\mathbf{k}$$

direction cosines are

$$\cos\alpha = \dfrac{1}{\sqrt{6}}$$

$$\cos\beta = \dfrac{1}{\sqrt{6}}$$

$$\cos\gamma = \dfrac{2}{\sqrt{6}}$$

VECTOR PRODUCT

The vector product **a x b** of 3D vectors **a** and **b** is a vector perpendicular to the plane containing **a** and **b**.

Since the resulting vector is a normal to the plane, it is often called **n** and has magnitude $|\mathbf{a}||\mathbf{b}|\sin\theta$

This gives

$$\mathbf{a} \times \mathbf{b} = \mathbf{n}|\mathbf{a}||\mathbf{b}|\sin\theta$$

The direction of **n** is found using the Right Hand Screw rule.

RIGHT HAND SCREW RULE.

If you curl your hand from **a** to **b**, your thumb will point towards **n**.
Looking at the unit vectors i,j,k

$$\mathbf{i} \times \mathbf{j} = \mathbf{k}|\mathbf{i}||\mathbf{j}|\sin\theta$$
$$= \mathbf{k}\sin 90°$$
$$= \mathbf{k}$$

also,

$$\mathbf{j} \times \mathbf{k} = \mathbf{i} \qquad \mathbf{k} \times \mathbf{j} = -\mathbf{i}$$
$$\mathbf{k} \times \mathbf{i} = \mathbf{j} \qquad \mathbf{i} \times \mathbf{k} = -\mathbf{j}$$
$$\mathbf{j} \times \mathbf{i} = -\mathbf{k}$$

$$\mathbf{i} \times \mathbf{i} = \mathbf{n}|\mathbf{i}||\mathbf{i}|\sin\theta$$
$$= \mathbf{n}\sin 0$$
$$= 0$$

$$\therefore \mathbf{i} \times \mathbf{i} = \mathbf{j} \times \mathbf{j} = \mathbf{k} \times \mathbf{k} = 0$$

OTHER PROPERTIES

$a \times b \neq b \times a$

$a \times b = -(b \times a)$

$b \times a = -a \times b$

So the vector product is not commutative

Vector **n** changes direction.

Given $a \neq 0$ and $b \neq 0$,

If $a \times b = 0$, then **a** and **b** have the same direction.

So $b = ka$ for some real number k

If k is real,

$(ka) \times b = a \times (kb) = k(a \times b)$

$a \times (b+c) = a \times b + a \times c$

$(a+b) \times c = a \times c + b \times c$

So the vector product is distributive

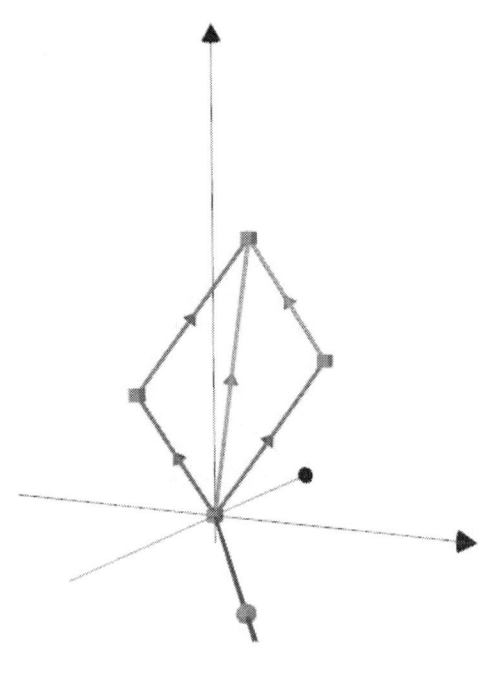

$|a \times b| = |a||b|\sin \theta$

describes the area of the parallelogram defined by **a** and **b**

COMPONENTS

Vectors **a** and **b** can be written in unit vector form as

$$\mathbf{a} = a_1\mathbf{i} + a_2\mathbf{j} + a_3\mathbf{k}$$
$$\mathbf{b} = b_1\mathbf{i} + b_2\mathbf{j} + b_3\mathbf{k}$$

$\mathbf{a} \times \mathbf{b}$ is the determinant of the matrix $\begin{pmatrix} i & j & k \\ a_1 & a_2 & a_3 \\ b_1 & b_2 & b_3 \end{pmatrix}$

$$\therefore \mathbf{a} \times \mathbf{b} = \begin{pmatrix} i & j & k \\ a_1 & a_2 & a_3 \\ b_1 & b_2 & b_3 \end{pmatrix}$$

$$= (a_2 b_3 - a_3 b_2)\mathbf{i} - (a_1 b_3 - a_3 b_1)\mathbf{j} + (a_1 b_2 - a_2 b_1)\mathbf{k}$$

which is why it is also know as the cross product.

$$\boxed{\begin{aligned} \mathbf{a} \times \mathbf{b} &= (a_2 b_3 - a_3 b_2)\mathbf{i} - (a_1 b_3 - a_3 b_1)\mathbf{j} + (a_1 b_2 - a_2 b_1)\mathbf{k} \\ &= (a_2 b_3 - a_3 b_2)\mathbf{i} + (a_3 b_1 - a_1 b_3)\mathbf{j} + (a_1 b_2 - a_2 b_1)\mathbf{k} \end{aligned}}$$

$$(\mathbf{a} \times \mathbf{b}) \cdot \mathbf{a} = \begin{pmatrix} a_2b_3 - a_3b_2 \\ a_3b_1 - a_1b_3 \\ a_1b_2 - a_2b_1 \end{pmatrix} \cdot \begin{pmatrix} a_1 \\ a_2 \\ a_3 \end{pmatrix}$$

$$= a_1(a_2b_3 - a_3b_2) + a_2(a_3b_1 - a_1b_3) + a_3(a_1b_2 - a_2b_1)$$
$$= a_1a_2b_3 - a_1a_3b_2 + a_2a_3b_1 - a_2a_1b_3 + a_3a_1b_2 - a_3a_2b_1$$
$$= 0$$

$$(\mathbf{a} \times \mathbf{b}) \cdot \mathbf{a} = (\mathbf{a} \times \mathbf{b}) \cdot \mathbf{b} = 0$$

To find a unit vector perpendicular to vectors **a** and **b**, calculate $n = \dfrac{\mathbf{a} \times \mathbf{b}}{|\mathbf{a} \times \mathbf{b}|}$

Example
Evaluate

a × **b**, where $\mathbf{a} = \begin{pmatrix} 4 \\ 5 \\ -2 \end{pmatrix}$ and $\mathbf{b} = \begin{pmatrix} 2 \\ -1 \\ -4 \end{pmatrix}$

$\mathbf{a} \times \mathbf{b} = (a_2b_3 - a_3b_2)\mathbf{i} + (a_3b_1 - a_1b_3)\mathbf{j} + (a_1b_2 - a_2b_1)\mathbf{k}$
$= (5 \times (-4) - (-2) \times (-1))\mathbf{i} + ((-2) \times 2 - 4 \times (-4))\mathbf{j} + (4 \times (-1) - 5 \times 2)\mathbf{k}$
$= (-20 - 2)\mathbf{i} + (-4 + 16)\mathbf{j} + (-4 - 10)\mathbf{k}$
$= -22\mathbf{i} + 12\mathbf{j} - 14\mathbf{k}$
$= \begin{pmatrix} -22 \\ 12 \\ -14 \end{pmatrix}$

> Note
>
> $(\mathbf{a} \times \mathbf{b}) \cdot \mathbf{a} = \begin{pmatrix} -22 \\ 12 \\ -14 \end{pmatrix} \cdot \begin{pmatrix} 4 \\ 5 \\ -2 \end{pmatrix}$
>
> $= -88 + 60 + 28 = 0$

Example

Calculate the distance from A(3,2,1) to the straight line passing through B(4,5,6) and C(7,8,9)

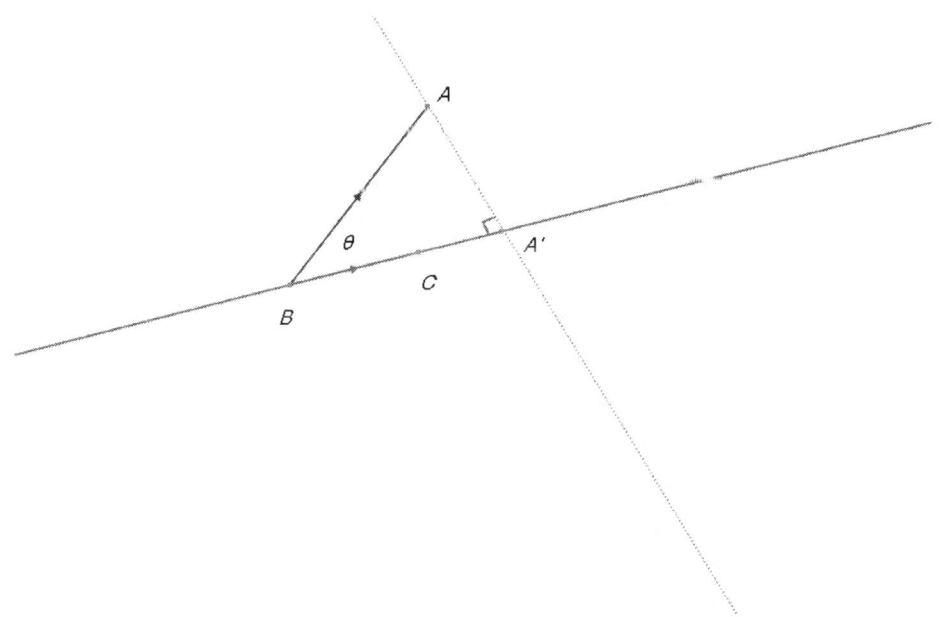

$$AA' = |\overrightarrow{BA}|\sin\theta$$

$$= \frac{|\overrightarrow{BC}||\overrightarrow{BA}|\sin\theta}{|\overrightarrow{BC}|}$$

$$= \frac{|\overrightarrow{BC} \times \overrightarrow{BA}|}{|\overrightarrow{BC}|}$$

$$\vec{BC} = \begin{pmatrix} 7 \\ 8 \\ 9 \end{pmatrix} - \begin{pmatrix} 4 \\ 5 \\ 6 \end{pmatrix} = \begin{pmatrix} 3 \\ 3 \\ 3 \end{pmatrix} \qquad \vec{BA} = \begin{pmatrix} 3 \\ 2 \\ 1 \end{pmatrix} - \begin{pmatrix} 4 \\ 5 \\ 6 \end{pmatrix} = \begin{pmatrix} -1 \\ -3 \\ -5 \end{pmatrix}$$

$$\therefore \frac{|\vec{BC} \times \vec{BA}|}{|\vec{BC}|} = \frac{\left| \begin{pmatrix} 3 \\ 3 \\ 3 \end{pmatrix} \times \begin{pmatrix} -1 \\ -3 \\ -5 \end{pmatrix} \right|}{\left| \begin{pmatrix} 3 \\ 3 \\ 3 \end{pmatrix} \right|}$$

$$= \frac{\left| \begin{pmatrix} i & j & k \\ 3 & 3 & 3 \\ -1 & -3 & -5 \end{pmatrix} \right|}{\sqrt{27}} = \frac{|-6\mathbf{i} + 12\mathbf{j} - 6\mathbf{k}|}{\sqrt{27}}$$

$$= \frac{\sqrt{216}}{\sqrt{27}} = \frac{6\sqrt{6}}{3\sqrt{3}} = \frac{2\sqrt{6}}{\sqrt{3}}$$

The point is $\dfrac{2\sqrt{6}}{\sqrt{3}}$ units from the line.

SCALAR TRIPLE PRODUCT

This describes the volume of a parallelepiped which has edges **a**, **b** and **c**.

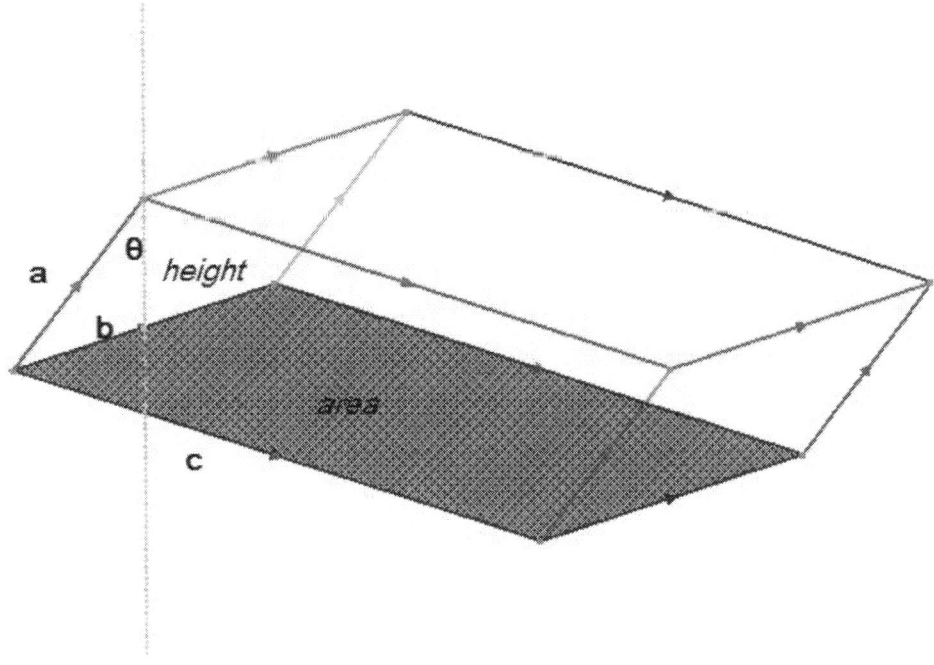

The base $|\mathbf{b}|$, $|\mathbf{c}|$ is separated by angle θ

The area of the base is $|\mathbf{b} \times \mathbf{c}|$

The height is $|\mathbf{a}| \cos \theta$

$\therefore V = Ah$

$= |\mathbf{b} \times \mathbf{c}| \cdot |\mathbf{a}| \cos \theta$

$= \mathbf{a}.(\mathbf{b} \times \mathbf{c})$

$$\mathbf{a}.(\mathbf{b} \times \mathbf{c}) = \begin{pmatrix} a_1 \\ a_2 \\ a_3 \end{pmatrix} \cdot \begin{pmatrix} b_2 c_3 - b_3 c_2 \\ b_3 c_1 - b_1 c_3 \\ b_1 c_2 - b_2 c_1 \end{pmatrix}$$

$$= a_1(b_2 c_3 - b_3 c_2) + a_2(b_3 c_1 - b_1 c_3) + a_3(b_1 c_2 - b_2 c_1)$$

$$(\mathbf{a} \times \mathbf{b}) \cdot \mathbf{c} = \begin{pmatrix} b_2c_3 - b_3c_2 \\ b_3c_1 - b_1c_3 \\ b_1c_2 - b_2c_1 \end{pmatrix} \cdot \begin{pmatrix} a_1 \\ a_2 \\ a_3 \end{pmatrix}$$

$$= a_1(b_2c_3 - b_3c_2) + a_2(b_3c_1 - b_1c_3) + a_3(b_1c_2 - b_2c_1)$$

$$\mathbf{a} \cdot (\mathbf{b} \times \mathbf{c}) = (\mathbf{a} \times \mathbf{b}) \cdot \mathbf{c}$$

Since **a.b** produces a scalar,

$(\mathbf{a.b}) \times \mathbf{c}$ is meaningless.

$\mathbf{a.b} \times \mathbf{c}$ means $\mathbf{a.(b \times c)}$

Example
Find the volume of the parallelepiped with vectors
a = **i** + 2**j** − 3**k**
b = 2**i** − 2**j** + **k**
c = **i** − 4**k**

$$\mathbf{a} \cdot (\mathbf{b} \times \mathbf{c}) = \begin{pmatrix} 1 \\ 2 \\ -3 \end{pmatrix} \cdot \begin{pmatrix} i & j & k \\ 2 & -2 & 1 \\ 1 & 0 & -4 \end{pmatrix}$$

$$\mathbf{a} \cdot (\mathbf{b} \times \mathbf{c}) = \begin{pmatrix} 1 \\ 2 \\ -3 \end{pmatrix} \cdot \begin{pmatrix} 8 - 0 \\ 1 - (-8) \\ 0 - (-2) \end{pmatrix}$$

$$= 8 + 2 \times 9 - 3 \times 2$$

$$= 20 \text{ units of volume}$$

PLANES

Let P(x,y,z) be a general point on plane π,

$\mathbf{n} = \begin{pmatrix} a \\ b \\ c \end{pmatrix}$ is a normal to π passing through point A.

P lies in π, so $\quad \overrightarrow{AP} \perp \mathbf{n}$
$\Rightarrow \quad \mathbf{n} \cdot \overrightarrow{AP} = 0$

Let $\mathbf{a} = \overrightarrow{OA}$ and $\mathbf{p} = \overrightarrow{OP}$, so $\overrightarrow{AP} = \overrightarrow{OP} - \overrightarrow{OA}$
$\Rightarrow \quad \mathbf{n} \cdot (\mathbf{p} - \mathbf{a}) = 0$
$\Rightarrow \quad \mathbf{n} \cdot \mathbf{p} - \mathbf{n} \cdot \mathbf{a} = 0$
$\Rightarrow \quad \mathbf{n} \cdot \mathbf{p} = \mathbf{n} \cdot \mathbf{a}$

This gives a vector equation for plane π

Since both **n** and A are fixed, **n**·**a** is a constant
Let $\mathbf{n} \cdot \mathbf{a} = k$
Then $\mathbf{n} \cdot \mathbf{p} - \mathbf{n} \cdot \mathbf{a} = 0$
$\Rightarrow \quad \mathbf{n} \cdot \mathbf{p} - k = 0$
$\Rightarrow \quad \mathbf{n} \cdot \mathbf{p} = k$

$\Rightarrow \begin{pmatrix} a \\ b \\ c \end{pmatrix} \cdot \begin{pmatrix} x \\ y \\ z \end{pmatrix} = k$

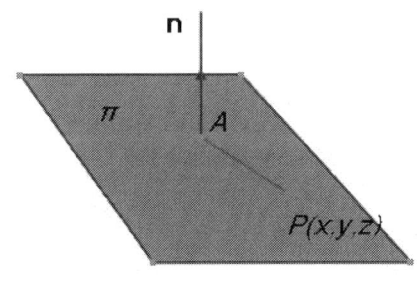

$\Rightarrow ax + by + cz = k$

Which gives a co-ordinate equation for plane π

> If $ax + by + cz = k$
> Then $\mathbf{n}=(a,b,c)$ is a normal vector

If P and Q lie on the plane,

$\mathbf{n} \cdot \mathbf{p} = k$ and $\mathbf{n} \cdot \mathbf{q} = k$
$\Rightarrow \mathbf{n} \cdot (\mathbf{p} - \mathbf{q}) = 0$
$\Rightarrow \mathbf{n}$ is orthogonal to all directions in the plane

Steps to find the equation of a plane

- Identify the normal.
(Use the cross product if necessary)
- The dot product of the normal and each point on the plane will be equal.
- Find the dot product of the normal and one given point on the plane.
- Use the result to find the equation of the plane.

Examples

Find the equation of the plane perpendicular to the vector $\mathbf{n}=2\mathbf{i}+3\mathbf{j}+6\mathbf{k}$ at the point G(1,2,0).
Does the point T(2,4,0) lie on the plane?

Let P (x,y,z) be a general point on the plane

so $\quad \overrightarrow{GP} \perp \mathbf{n}$

$\Rightarrow \quad \mathbf{n} \cdot \overrightarrow{GP} = 0$

Let $\mathbf{g} = \overrightarrow{OG}$ and $\mathbf{p} = \overrightarrow{OP}$, so $\overrightarrow{GP} = \overrightarrow{OP} - \overrightarrow{OG}$

$$\mathbf{n} \cdot \overrightarrow{GP} = 0$$
$$\mathbf{n} \cdot (\mathbf{p} - \mathbf{g}) = 0$$
$$\Rightarrow \mathbf{n} \cdot \mathbf{p} - \mathbf{n} \cdot \mathbf{g} = 0$$

Let $\mathbf{n} \cdot \mathbf{g} = k$

$\Rightarrow \begin{pmatrix} 2 \\ 3 \\ 6 \end{pmatrix} \cdot \begin{pmatrix} 1 \\ 2 \\ 0 \end{pmatrix} = k$

$\Rightarrow 2 + 6 = k$

$\Rightarrow 8 = k$

Then $\mathbf{n} \cdot \mathbf{p} - \mathbf{n} \cdot \mathbf{g} = 0$

$\Rightarrow \mathbf{n} \cdot \mathbf{p} - 8 = 0$

$\Rightarrow \mathbf{n} \cdot \mathbf{p} = 8$

$\Rightarrow \begin{pmatrix} 2 \\ 3 \\ 6 \end{pmatrix} \cdot \begin{pmatrix} x \\ y \\ z \end{pmatrix} = 8$

$2x + 3y + 6z = 8$

The plane has equation $2x + 3y + 6z = 8$

Substitute T(2,4,0) into $2x + 3y + 6z = 8$

$2 \times 2 + 3 \times 4 + 6 \times 0 = 4 + 12 = 16$

since $16 \neq 8$, T does not lie on the plane.

Given R(1,2,3) and S(-2,-1,-3), find the equation of the plane passing through R which is perpendicular to RS.

Let P (x,y,z) be a general point on the plane
P lies in π, so $\quad \overrightarrow{RP} \perp \mathbf{n}$
$$\Rightarrow \mathbf{n} \cdot \overrightarrow{RP} = 0$$

Let $\mathbf{r} = \overrightarrow{OR}$ and $\mathbf{p} = \overrightarrow{OP}$, so $\overrightarrow{RP} = \overrightarrow{OP} - \overrightarrow{OR}$

$$\mathbf{n} \cdot \overrightarrow{RP} = 0$$
$$\mathbf{n} \cdot (\mathbf{p} - \mathbf{r}) = 0$$
$$\Rightarrow \mathbf{n} \cdot \mathbf{p} - \mathbf{n} \cdot \mathbf{r} = 0$$

Here, the line $\overrightarrow{RS}$ describes the normal, $\mathbf{n}$
$$\Rightarrow \overrightarrow{RS} \cdot \mathbf{p} - \overrightarrow{RS} \cdot \mathbf{r} = 0$$
$$\Rightarrow \overrightarrow{RS} \cdot \mathbf{p} = \overrightarrow{RS} \cdot \mathbf{r}$$

$$\overrightarrow{RS} = \begin{pmatrix} -2 \\ -1 \\ -3 \end{pmatrix} - \begin{pmatrix} 1 \\ 2 \\ 3 \end{pmatrix} = \begin{pmatrix} -3 \\ -3 \\ -6 \end{pmatrix}$$

$$\Rightarrow \overrightarrow{RS} \cdot \mathbf{r} = \begin{pmatrix} -3 \\ -3 \\ -6 \end{pmatrix} \cdot \begin{pmatrix} 1 \\ 2 \\ 3 \end{pmatrix}$$
$$= -3 - 6 - 18 = -27$$

$$\overrightarrow{RS} \cdot \mathbf{p} = -27$$
$$\Rightarrow \begin{pmatrix} -3 \\ -3 \\ -6 \end{pmatrix} \cdot \begin{pmatrix} x \\ y \\ z \end{pmatrix} = -27$$
$$\Rightarrow -3x - 3y - 6z = -27$$
$$\Rightarrow x + y + 2z = 9$$
is an equation of the plane.

Find the equation of the plane passing through A(1,3,1), B(2,0,-2) and C(4,5,6)

$$\vec{AB} = \begin{pmatrix} 2 \\ 0 \\ -2 \end{pmatrix} - \begin{pmatrix} 1 \\ 3 \\ 1 \end{pmatrix} = \begin{pmatrix} 1 \\ -3 \\ -3 \end{pmatrix} \qquad \vec{AC} = \begin{pmatrix} 4 \\ 5 \\ 6 \end{pmatrix} - \begin{pmatrix} 1 \\ 3 \\ 1 \end{pmatrix} = \begin{pmatrix} 3 \\ 2 \\ 5 \end{pmatrix}$$

Let P (x,y,z) be a general point on the plane

$$\mathbf{n} \cdot \vec{AP} = 0$$

Let $\mathbf{a} = \vec{OA}$ and $\mathbf{p} = \vec{OP}$, so $\vec{AP} = \vec{OP} - \vec{OA}$

$\Rightarrow \mathbf{n} \cdot (\mathbf{p} - \mathbf{a}) = 0$

Here, $\mathbf{n} = |\vec{AB} \times \vec{AC}|$

$$\mathbf{n} = |\vec{AB} \times \vec{AC}| = \begin{vmatrix} i & j & k \\ 1 & -3 & -3 \\ 3 & 2 & 5 \end{vmatrix}$$

$$= -9\mathbf{i} - 14\mathbf{j} + 11\mathbf{k}$$

$$\mathbf{p} - \mathbf{a} = \begin{pmatrix} x \\ y \\ z \end{pmatrix} - \begin{pmatrix} 1 \\ 3 \\ 1 \end{pmatrix} = \begin{pmatrix} x-1 \\ y-3 \\ z-1 \end{pmatrix}$$

$\Rightarrow \mathbf{n} \cdot \vec{AP} = 0$

$$\Rightarrow \begin{pmatrix} -9 \\ -14 \\ 11 \end{pmatrix} \cdot \begin{pmatrix} x-1 \\ y-3 \\ z-1 \end{pmatrix} = 0$$

$\Rightarrow -9(x-1) - 14(y-3) + 11(z-1) = 0$
$\Rightarrow -9x + 9 - 14y + 42 + 11z - 11 = 0$
$\Rightarrow -9x - 14y + 11z = -40$

PLANES IN SPACE

A plane in space can be found if:-
- 3 points on the plane are known

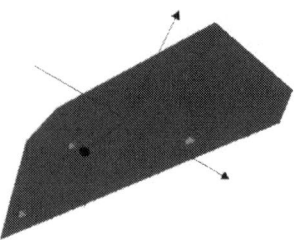

- 2 non-parallel intersecting lines on the plane are known

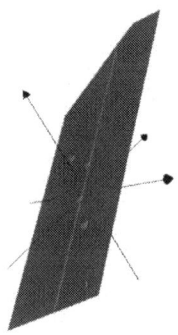

- One point on the plane and a normal to the plane are known

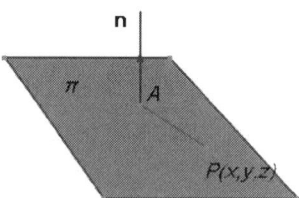

THE ANGLE BETWEEN TWO PLANES

The angle between two planes is found using the scalar product.

It is equal to the acute angle determined by the normal vectors of the planes.

Example

Calculate the angle between the planes
π_1: $x + 2y - 2z = 5$
and π_2: $6x - 3y + 2z = 8$

let $\mathbf{a} = \begin{pmatrix} 1 \\ 2 \\ -2 \end{pmatrix}$ represent the normal for π_1

and $\mathbf{b} = \begin{pmatrix} 6 \\ -3 \\ 2 \end{pmatrix}$ represent the normal for π_2

$|\mathbf{a}| = \sqrt{1+4+4}$
$= 3$

$|\mathbf{b}| = \sqrt{36+9+4}$
$= 7$

$\cos\theta = \dfrac{a_1 b_1 + a_2 b_2 + a_3 b_3}{|\mathbf{a}||\mathbf{b}|}$

$\Rightarrow \cos\theta = \dfrac{1 \times 6 - 2 \times 3 - 2 \times 2}{21}$

$\Rightarrow \cos\theta = \dfrac{-4}{21}$

$\Rightarrow \theta = 100.98°$ i.e obtuse

$\Rightarrow \theta = 79.02°$

THE DISTANCE BETWEEN PARALLEL PLANES

Let P be a point on plane π_1 : $ax + by + cz = n$
and Q be a point on plane π_2 : $ax + by + cz = m$
So **a.x** = n   and   **a.x** = m

Since the planes are parallel, they share the common normal, **a**
where **a** = (a**i** + b**j** + c**k**)
The distance between the planes is

$$\boxed{PQ = \frac{|m-n|}{|\mathbf{a}|}}$$

Example
Calculate the distance between the planes
π_1: $x + 2y - 2z = 5$
and π_2: $6x + 12y - 12z = 8$

$x + 2y - 2z = 5$

$x + 2y - 2z = 5$

so $\mathbf{a} = \begin{pmatrix} 1 \\ 2 \\ -2 \end{pmatrix}$, $n = 5$

$6x + 12y - 12z = 8$

$x + 2y - 2z = \dfrac{4}{3}$

and $m = \dfrac{4}{3}$

$PQ = \dfrac{|m-n|}{|\mathbf{a}|}$

$= \dfrac{\left|\dfrac{4}{3} - 5\right|}{\sqrt{1+4+4}}$

$= \dfrac{\dfrac{11}{3}}{3}$

$= \dfrac{11}{9} = 1\dfrac{2}{9}$ units

COPLANAR VECTORS

If a relationship exists between the vectors **a**, **b** and **c** such that $\mathbf{c}=\lambda\mathbf{a}+\mu\mathbf{b}$

then a, b and c are co-planar.

If three vectors are co-planar,

$$\boxed{\mathbf{c} = \lambda\mathbf{a} + \mu\mathbf{b}}$$

VECTOR EQUATION OF A PLANE

If **a, b** and **c** are position vectors on a plane

$\boxed{\mathbf{r} = \mathbf{a} + \lambda\mathbf{b} + \mu\mathbf{c}}$ is the vector equation of the plane.

A is a point on the plane, **b** and **c** are vectors parallel to the plane.

When the position vectors are used,

$\boxed{\mathbf{r} = (1-\lambda-u)\mathbf{a} + \lambda\mathbf{b} + \mu\mathbf{c}}$ is the vector equation of the plane.

Example

Find a vector equation of the plane through the points
A(-1,-2,-3) , B(-2,0,1) and C(-4,-1,-1)

$\mathbf{r} = (1-\lambda-\mu)\mathbf{a} + \lambda\mathbf{b} + \mu\mathbf{c}$

$= (1-\lambda-\mu)\begin{pmatrix}-1\\-2\\-3\end{pmatrix} + \lambda\begin{pmatrix}-2\\0\\1\end{pmatrix} + \mu\begin{pmatrix}-4\\-1\\-1\end{pmatrix}$

$= \begin{pmatrix}-(1-\lambda-\mu)-2\lambda-4\mu\\-2(1-\lambda-\mu)-\mu\\-3(1-\lambda-\mu)+\lambda-\mu\end{pmatrix}$

$= \begin{pmatrix}-1+\lambda+\mu-2\lambda-4\mu\\-2+2\lambda+2\mu-\mu\\-3+3\lambda+3\mu+\lambda-\mu\end{pmatrix}$

$= \begin{pmatrix}-1-\lambda-3\mu\\-2+2\lambda+\mu\\-3+4\lambda+2\mu\end{pmatrix}$

$= (-1-\lambda-3\mu)\mathbf{i} + (-2+2\lambda+\mu)\mathbf{j} + (-3+4\lambda+2\mu)\mathbf{k}$

THE EQUATIONS OF A LINE

A line can be described when a point on it and its direction vector – a vector parallel to the line – are known.

In the diagram below, the line L passes through points $A(x_1,y_1,z_1)$ and $P(x,y,z)$.

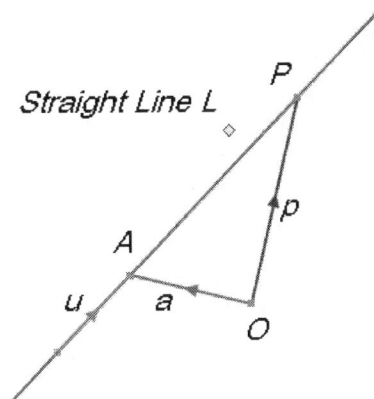

u is the direction vector **ai +bj +ck**
Being on the line, it has the same direction as any parallel line.
O is the origin.
a and **p** represent the position vectors of A and P.

P is on line L
$\Rightarrow \overrightarrow{AP} = \lambda \mathbf{u}$ for some scalar λ
$\Rightarrow \mathbf{p - a} = \lambda \mathbf{u}$
$\Rightarrow \mathbf{p} = \mathbf{a} + \lambda \mathbf{u}$

> **p = a + λu**
> is the vector equation of the line
> convention often replaces **p** with **r**

$$\mathbf{r} = \mathbf{a} + \lambda \mathbf{u}$$

If two points are known, say A and B then $\mathbf{u} = \overrightarrow{AB} = \mathbf{b} - \mathbf{a}$

$\Rightarrow \quad \mathbf{r} = \mathbf{a} + \lambda(\mathbf{b} - \mathbf{a})$

$\Rightarrow \quad \mathbf{r} = \mathbf{a} + \lambda\mathbf{b} - \lambda\mathbf{a}$

$\Rightarrow \quad \mathbf{r} = (1 - \lambda)\mathbf{a} + \lambda\mathbf{b}$

In component form, $\mathbf{r} = \mathbf{a} + \lambda\mathbf{u}$ becomes

$$\begin{pmatrix} x \\ y \\ z \end{pmatrix} = \begin{pmatrix} x_1 \\ y_1 \\ z_1 \end{pmatrix} + \lambda \begin{pmatrix} a \\ b \\ c \end{pmatrix}$$

Thus

$$\begin{pmatrix} x \\ y \\ z \end{pmatrix} = \begin{pmatrix} x_1 + \lambda a \\ y_1 + \lambda b \\ z_1 + \lambda c \end{pmatrix}$$

giving the parametric equations

$$x = x_1 + \lambda a, \quad y = y_1 + \lambda b, \quad z = z_1 + \lambda c$$

so

$$\frac{x - x_1}{a} = \lambda \quad \frac{y - y_1}{b} = \lambda \quad \frac{z - z_1}{c} = \lambda$$

Giving the symmetric form

$$\frac{x - x_1}{a} = \frac{y - y_1}{b} = \frac{z - z_1}{c} = \lambda$$

(aka standard form, canonical form, co-ordinate equation)

Examples

Find the vector equation of the straight line through (3,2,1) which is parallel to the vector 2**i** +3**j** +4**k**

$\mathbf{r} = \mathbf{a} + \lambda\mathbf{u}$
$\Rightarrow \mathbf{r} = 3\mathbf{i} + 2\mathbf{j} + \mathbf{k} + \lambda(2\mathbf{i} + 3\mathbf{j} + 4\mathbf{k})$

$$\Rightarrow \mathbf{r} = \begin{pmatrix} 3 \\ 2 \\ 1 \end{pmatrix} + \lambda \begin{pmatrix} 2 \\ 3 \\ 4 \end{pmatrix}$$

are the vector equations of the line

Find the vector form of the equation of the straight line which has parametric equations

$x = 4 - 2\lambda \qquad y = 7 + \lambda \qquad z = 3 - 4\lambda$

$$\begin{pmatrix} x \\ y \\ z \end{pmatrix} = \begin{pmatrix} 4 \\ 7 \\ 3 \end{pmatrix} + \lambda \begin{pmatrix} -2 \\ 1 \\ -4 \end{pmatrix}$$

$\Rightarrow \mathbf{r} = 4\mathbf{i} + 7\mathbf{j} + 3\mathbf{k} + \lambda(-2\mathbf{i} + \mathbf{j} - 4\mathbf{k})$

Find the Cartesian form of the line which has position vector $3\mathbf{i} + 2\mathbf{j} + \mathbf{k}$ and is parallel to the vector $\mathbf{i} - \mathbf{j} + \mathbf{k}$

$$\mathbf{r} = 3\mathbf{i} + 2\mathbf{j} + \mathbf{k} + \lambda(\mathbf{i} - \mathbf{j} + \mathbf{k})$$

$$\Rightarrow \begin{pmatrix} x \\ y \\ z \end{pmatrix} = \begin{pmatrix} 3 \\ 2 \\ 1 \end{pmatrix} + \lambda \begin{pmatrix} 1 \\ -1 \\ 1 \end{pmatrix}$$

$\therefore x = 3 + \lambda \qquad y = 2 - \lambda \qquad z = 1 + \lambda$

$$\frac{x-3}{1} = \frac{y-2}{-1} = \frac{z-1}{1} = \lambda$$

$\Rightarrow \quad x - 3 = 2 - y = z - 1 = \lambda$

Find the vector equation of the line passing through A(1,2,3) and B(4,5,6)

$\mathbf{r} = \mathbf{a} + \lambda \mathbf{u}$
$\mathbf{a} = \mathbf{i} + 2\mathbf{j} + 3\mathbf{k}$
$\mathbf{b} = 4\mathbf{i} + 5\mathbf{j} + 6\mathbf{k}$
$\mathbf{u} = \overrightarrow{AB} = \mathbf{b} - \mathbf{a}$
$\Rightarrow \mathbf{u} = 3\mathbf{i} + 3\mathbf{j} + 3\mathbf{k}$
$\Rightarrow \mathbf{r} = \mathbf{i} + 2\mathbf{j} + 3\mathbf{k} + \lambda(3\mathbf{i} + 3\mathbf{j} + 3\mathbf{k})$

---

alternatively
$\mathbf{r} = (1-\lambda)\mathbf{a} + \lambda\mathbf{b}$
$\Rightarrow \mathbf{r} = (1-\lambda)(\mathbf{i} + 2\mathbf{j} + 3\mathbf{k}) + \lambda(4\mathbf{i} + 5\mathbf{j} + 6\mathbf{k})$
$\Rightarrow \mathbf{r} = (\mathbf{i} + 2\mathbf{j} + 3\mathbf{k}) - \lambda(\mathbf{i} + 2\mathbf{j} + 3\mathbf{k}) + \lambda(4\mathbf{i} + 5\mathbf{j} + 6\mathbf{k})$
$\Rightarrow \mathbf{r} = (\mathbf{i} + 2\mathbf{j} + 3\mathbf{k}) + \lambda(3\mathbf{i} + 3\mathbf{j} + 3\mathbf{k})$

A line L has equations

$$\frac{x+2}{3} = \frac{y-1}{2} = \frac{3-z}{4}$$

Is the vector $\mathbf{s} = 6\mathbf{i} + 4\mathbf{j} - 8\mathbf{k}$ parallel to L ?

$$\frac{x+2}{3} = \frac{y-1}{2} = \frac{3-z}{4} = \lambda$$

$\Rightarrow x = -2 + 3\lambda \qquad y = 1 + 2\lambda \qquad z = 3 - 4\lambda$

$\Rightarrow \mathbf{r} = \begin{pmatrix} -2 \\ 1 \\ 3 \end{pmatrix} + \lambda \begin{pmatrix} 3 \\ 2 \\ -4 \end{pmatrix}$

$\mathbf{r} = \mathbf{a} + \lambda \mathbf{u}$

$\Rightarrow$ (-2,1,3) is a point on L
and $\lambda(3\mathbf{i} + 2\mathbf{j} - 4\mathbf{k})$ is a direction vector.

$\mathbf{s}$ has direction ratio $6 : 4 : -8 = 3 : 2 : -4$

The direction ratios of $\mathbf{s}$ and $\mathbf{u}$ are the same
$\Rightarrow \mathbf{s} \parallel \mathbf{u}$

THE ANGLE BETWEEN A LINE AND A PLANE

The angle θ between a line and a plane is the complement of the angle between the line and the normal to the plane.

If the line has direction vector **u** and the normal to the plane is **a**, then

$$\sin\theta° = |\cos(90-\theta)°| = \frac{|\mathbf{a}\cdot\mathbf{u}|}{|\mathbf{a}||\mathbf{u}|}$$

Example

Given the equations

$$\frac{x-4}{3}=\frac{y-3}{2}=\frac{z-5}{6}$$

and the plane $6x+3y-2z=14$

1) Find the point of intersection
2) Find the angle the line makes with the plane.

$$\frac{x-4}{3}=\frac{y-3}{2}=\frac{z-5}{6}=\lambda$$

$\Rightarrow\ x=4+3\lambda \qquad y=3+2\lambda \qquad z=5+6\lambda$

$\therefore (4+3\lambda,\ 3+2\lambda,5+6\lambda)$ lies on the plane

$\Rightarrow 6(4+3\lambda)+3(3+2\lambda)-2(5+6\lambda)=14$
$\Rightarrow 24+18\lambda+9+6\lambda-10-12\lambda=14$
$\Rightarrow 23+12\lambda=14$
$\Rightarrow \lambda=\dfrac{14-23}{12}=\dfrac{-3}{4}$

$$x = 4 + 3 \times \frac{-3}{4} \qquad y = 3 + 2 \times \frac{-3}{4} \qquad z = 5 + 6 \times \frac{-3}{4}$$

$$\Rightarrow x = \frac{16-9}{4} \qquad y = \frac{12-6}{4} \qquad z = \frac{20-18}{4}$$

$$\Rightarrow x = \frac{7}{4} \qquad y - \frac{3}{2} \qquad z = \frac{1}{2}$$

The point of intersection is $\left(\frac{7}{4}, \frac{3}{2}, \frac{1}{2}\right)$

$$\sin \theta° = |\cos(90-\theta)°| = \frac{|\mathbf{a} \cdot \mathbf{u}|}{|\mathbf{a}||\mathbf{u}|}$$

$\mathbf{a} = 6\mathbf{i} + 3\mathbf{j} - 2\mathbf{k}$
$\mathbf{u} = 3\mathbf{i} + 2\mathbf{j} + 6\mathbf{k}$

$$\sin \theta° = \frac{|(6\mathbf{i} + 3\mathbf{j} - 2\mathbf{k}) \cdot (3\mathbf{i} + 2\mathbf{j} + 6\mathbf{k})|}{\left(\left|\sqrt{36+9+4}\right|\right)\left|\sqrt{9+4+36}\right|}$$

$\Rightarrow \sin \theta° = \frac{12}{49} \quad (0 \leq \theta \leq 90)$

$\Rightarrow \theta = 14.175°$

The angle of intersection is $14.2°$

THE INTERSECTION OF TWO LINES

Example

Show that the lines with equations

$$\mathbf{r} = \begin{pmatrix} 3 \\ 4 \\ 1 \end{pmatrix} + \lambda_1 \begin{pmatrix} 4 \\ 1 \\ 0 \end{pmatrix}$$

and $\dfrac{x+1}{12} = \dfrac{y-7}{6} = \dfrac{z-5}{3} = \lambda_2$ intersect,

find the point of intersection and the equation of the plane containing the lines.

$$\mathbf{r} = \begin{pmatrix} 3 \\ 4 \\ 1 \end{pmatrix} + \lambda_1 \begin{pmatrix} 4 \\ 1 \\ 0 \end{pmatrix}$$

$\Rightarrow$ $x = 3 + 4\lambda_1$ $y = 4 + \lambda_1$ $z = 1$

and

$$\dfrac{x+1}{12} = \dfrac{y-7}{6} = \dfrac{z-5}{3} = \lambda_2$$

$\Rightarrow$ $x = -1 + 12\lambda_2$ $y = 7 + 6\lambda_2$ $z = 5 + 3\lambda_2$

Equating co-ordinates

$3 + 4\lambda_1 = -1 + 12\lambda_2 \qquad 4 + \lambda_1 = 7 + 6\lambda_2 \qquad 1 = 5 + 3\lambda_2$

$4\lambda_1 = -4 + 12\lambda_2 \qquad (1)$
$\lambda_1 = 3 + 6\lambda_2 \qquad (2)$
$0 = 4 + 3\lambda_2 \qquad (3)$

From (3), $\quad 3\lambda_2 = -4$

$\Rightarrow \lambda_2 = \dfrac{-4}{3}$

$\Rightarrow \lambda_1 = 3 + 6 \times \dfrac{-4}{3} = -5$

substituting

$\begin{aligned} x &= 3 + 4\lambda_1 & y &= 4 + \lambda_1 & z &= 1 \\ &= 3 - 20 & &= 4 - 5 \\ &= -17 & &= -1 \end{aligned}$

Intersection point is $(-17, -1, 1)$

Let $A(-17,-1,1)$ $B(3,4,1)$ $C(-1,7,5)$ be the points from the lines above

$\overrightarrow{AB} = \begin{pmatrix} 3 \\ 4 \\ 1 \end{pmatrix} - \begin{pmatrix} -17 \\ -1 \\ 1 \end{pmatrix} = \begin{pmatrix} 20 \\ 5 \\ 0 \end{pmatrix}$

$\overrightarrow{AC} = \begin{pmatrix} -1 \\ 7 \\ 5 \end{pmatrix} - \begin{pmatrix} -17 \\ -1 \\ 1 \end{pmatrix} = \begin{pmatrix} 16 \\ 8 \\ 4 \end{pmatrix}$

$\mathbf{n} \cdot \overrightarrow{AP} = 0$

Let $\mathbf{a} = \overrightarrow{OA}$ and $\mathbf{p} = \overrightarrow{OP}$, so $\overrightarrow{AP} = \overrightarrow{OP} - \overrightarrow{OA}$

$\Rightarrow \quad \mathbf{n} \cdot (\mathbf{p} - \mathbf{a}) = 0$

Here, $\mathbf{n} = \left|\overrightarrow{AB} \times \overrightarrow{AC}\right|$

$$\mathbf{n} = \left|\overrightarrow{AB} \times \overrightarrow{AC}\right| = \begin{vmatrix} i & j & k \\ 20 & 5 & 0 \\ 16 & 8 & 4 \end{vmatrix}$$

$$= 20\mathbf{i} - 80\mathbf{j} + 80\mathbf{k}$$
$$= \mathbf{i} - 4\mathbf{j} + 4\mathbf{k}$$

$$\mathbf{p} \cdot \mathbf{a} = \begin{pmatrix} x \\ y \\ z \end{pmatrix} - \begin{pmatrix} -17 \\ -1 \\ 1 \end{pmatrix} = \begin{pmatrix} x+17 \\ y+1 \\ z-1 \end{pmatrix}$$

$\Rightarrow \mathbf{n} \cdot \overrightarrow{AP} = 0$

$$\Rightarrow \begin{pmatrix} 1 \\ -4 \\ 4 \end{pmatrix} \cdot \begin{pmatrix} x+17 \\ y+1 \\ z-1 \end{pmatrix} = 0$$

$\Rightarrow x + 17 - 4(y+1) + 4(z-1) = 0$
$\Rightarrow x + 17 - 4y - 4 + 4z - 4 = 0$
$\Rightarrow x - 4y + 4z + 9 = 0$

THE INTERSECTION OF TWO PLANES

To find the equations of the line of intersection of two planes, a direction vector and point on the line is required.

Since the line of intersection lies in both planes, the direction vector is parallel to the vector products of the normal of each plane.

Example
Find the equation for the line of intersection of the planes
$-3x + 2y + z = -5$

$7x + 3y - 2z = -2$

$-3x + 2y + z = -5 \qquad\qquad 7x + 3y - 2z = -2$

Let $z = 0$

Then $\qquad -3x + 2y = -5 \dots (1)$

and $\qquad 7x + 3y = -2 \dots (2)$

$(2) \times 2 \qquad 14x + 6y = -4$

$(1) \times -3 \qquad \underline{9x - 6y = 15}$

add $\qquad\qquad 23x = 11$

$\Rightarrow \qquad x = \dfrac{11}{23}$

subst in (1)

$$-\dfrac{33}{23} + 2y = -5$$

$\Rightarrow \qquad y = \dfrac{-5 + \dfrac{33}{23}}{2} = \dfrac{-41}{23}$

The point $\left(\dfrac{11}{23}, \dfrac{-41}{23}, 0\right)$ is on the line of intersection

Normal vectors are $\mathbf{u} = -3\mathbf{i} + 2\mathbf{j} + \mathbf{k}$
and $\mathbf{v} = 7\mathbf{i} + 3\mathbf{j} - 2\mathbf{k}$

$$\mathbf{u} \times \mathbf{v} = \begin{vmatrix} i & j & k \\ -3 & 2 & 1 \\ 7 & 3 & -2 \end{vmatrix}$$
$$= -7\mathbf{i} + \mathbf{j} - 23\mathbf{k}$$

$$\mathbf{r} = \begin{pmatrix} \frac{11}{23} \\ \frac{-41}{23} \\ 0 \end{pmatrix} + \lambda_1 \begin{pmatrix} -7 \\ 1 \\ -23 \end{pmatrix}$$

$$= \begin{pmatrix} \frac{11}{23} \\ \frac{-41}{23} \\ 0 \end{pmatrix} + \lambda_1 \begin{pmatrix} \frac{7}{23} \\ \frac{-1}{23} \\ 1 \end{pmatrix}$$

$\Rightarrow \quad x = \dfrac{11}{23} + \dfrac{7}{23}\lambda_1 \qquad y = \dfrac{-41}{23} - \dfrac{1}{23}\lambda_1 \qquad z = \lambda_1$

THE DISTANCE FROM A POINT TO A PLANE

To find the distance of a point P to a plane

1) Find the equation of the projection PP' by using the normal to the plane and the point P.

2) Find the co-ordinates of P', the intersection with the plane.

3) Apply the distance formula to PP'

Alternatively

> The distance D between a point $P_0(x_0, y_0, z_0)$ and the plane $ax + by + cz + d = 0$ is
>
> $$D = \frac{|ax_0 + by_0 + cz_0 + d|}{\sqrt{a^2 + b^2 + c^2}}$$

Example

Find the distance between the point (3,1,-2) and the plane
x +2y +2z = -4

$$\mathbf{r} = \mathbf{u} + \lambda_1 \begin{pmatrix} 1 \\ 2 \\ 2 \end{pmatrix}$$

$$\mathbf{r} = \begin{pmatrix} 3 \\ 1 \\ -2 \end{pmatrix} + \lambda_1 \begin{pmatrix} 1 \\ 2 \\ 2 \end{pmatrix}$$

$$\Rightarrow \quad x = 3 + \lambda_1 \qquad y = 1 + 2\lambda_1 \qquad z = -2 + 2\lambda_1$$

Plane equation is $x + 2y + 2z + 4 = 0$

$$\Rightarrow 3 + \lambda_1 + 2(1 + 2\lambda_1) + 2(-2 + 2\lambda_1) + 4 = 0$$
$$\Rightarrow 3 + \lambda_1 + 2 + 4\lambda_1 - 4 + 4\lambda_1 + 4 = 0$$
$$\Rightarrow 5 + 9\lambda_1 = 0$$
$$\Rightarrow \lambda_1 = \frac{-5}{9}$$

$$\Rightarrow \quad x = 3 - \frac{5}{9} \qquad y = 1 - \frac{10}{9} \qquad z = -2 - \frac{10}{9}$$

$$P'\left(\frac{22}{9}, -\frac{1}{9}, -\frac{28}{9}\right)$$

$$PP' = \begin{pmatrix} -\frac{5}{9} \\ -\frac{10}{9} \\ -\frac{10}{9} \end{pmatrix} = \frac{-5}{9} \begin{pmatrix} 1 \\ 2 \\ 2 \end{pmatrix}$$

$$\Rightarrow |PP'| = \left| \frac{-5}{9} \sqrt{1+4+4} \right|$$

$$= \left| \frac{-5}{3} \right|$$

$$= \frac{5}{3} \text{ units}$$

Alternatively

$x + 2y + 2z = -4$ at $(3, 1, -2)$
$\Rightarrow x + 2y + 2z + 4 = 0$

$$D = \frac{|ax_0 + by_0 + cz_0 + d|}{\sqrt{a^2 + b^2 + c^2}}$$

$$= \frac{|3 + 2 - 4 + 4|}{\sqrt{1 + 4 + 4}}$$

$$= \frac{5}{3}$$

The distance is $\frac{5}{3}$ units

THE DISTANCE FROM A POINT TO A LINE

To find the distance of a point P to a Line L:-

1) Let the line have direction vector **u** and parameter λ

2) Find the co-ordinates of PP' by using the scalar product with u and the point P.

3) Apply the distance formula to PP'

Example

Find the distance between the line

$$\frac{x+3}{-6} = \frac{y-2}{9} = \frac{z+8}{6}$$

and the point P $(-1, 7, 4)$

$$P' = \begin{pmatrix} -3 \\ 2 \\ -8 \end{pmatrix} + \lambda_1 \begin{pmatrix} -6 \\ 9 \\ 6 \end{pmatrix}$$

$\Rightarrow \quad x = -3 - 6\lambda_1 \qquad y = 2 + 9\lambda_1 \qquad z = -8 + 6\lambda_1$

$P'(-3-6\lambda_1, \ 2+9\lambda_1, \ -8+6\lambda_1)$

$$\overrightarrow{PP'} = \begin{pmatrix} -3-6\lambda \\ 2+9\lambda_1 \\ -8+6\lambda_1 \end{pmatrix} - \begin{pmatrix} -1 \\ 7 \\ 4 \end{pmatrix} = \begin{pmatrix} -2-6\lambda \\ -5+9\lambda \\ -12+6\lambda \end{pmatrix}$$

$\overrightarrow{PP'} \cdot \mathbf{u} = 0$

$$\Rightarrow \quad \begin{pmatrix} -2-6\lambda \\ -5+9\lambda \\ -12+6\lambda \end{pmatrix} \cdot \begin{pmatrix} -6 \\ 9 \\ 6 \end{pmatrix} = 0$$

$\Rightarrow -6(-2-6\lambda) + 9(-5+9\lambda) + 6(-12+6\lambda) = 0$

$$\Rightarrow 12 + 36\lambda - 45 + 81\lambda - 72 + 36\lambda = 0$$
$$\Rightarrow -105 + 153\lambda = 0$$
$$\Rightarrow \lambda = \frac{105}{153} = \frac{35}{51}$$

$$\overrightarrow{PP'} = \begin{pmatrix} -2-6\lambda \\ -5+9\lambda \\ -12+6\lambda \end{pmatrix} = \begin{pmatrix} -2-6\times\frac{35}{51} \\ -5+9\times\frac{35}{51} \\ -12+6\times\frac{35}{51} \end{pmatrix} = \begin{pmatrix} \frac{-104}{17} \\ \frac{20}{17} \\ \frac{-134}{17} \end{pmatrix} = \frac{1}{17}\begin{pmatrix} -104 \\ 20 \\ -134 \end{pmatrix}$$

$$\Rightarrow PP' = \frac{1}{17}\sqrt{29172} = 10.04$$

The distance is 10.04 units

THE INTERSECTION OF THREE PLANES

To solve the intersection, use the equations of the plane
ax +by +cz +d = 0 to form an augmented matrix, which is solved for x, y and z.

The intersection between three planes could be

- **A single point** (A unique solution is found)

Example

$x + y + z = 2$
$4x + 2y + z = 4$
$x - y + z = 4$

$$\begin{pmatrix} 1 & 1 & 1 & 2 \\ 4 & 2 & 1 & 4 \\ 1 & -1 & 1 & 4 \end{pmatrix} \rightarrow \begin{pmatrix} 1 & 1 & 1 & 2 \\ 0 & 1 & 0 & -1 \\ 0 & 0 & 1 & 2 \end{pmatrix} \rightarrow \begin{pmatrix} 1 & 0 & 0 & 1 \\ 0 & 1 & 0 & -1 \\ 0 & 0 & 1 & 2 \end{pmatrix}$$

Point $(1, -1, 2)$

- **A line of intersection** (An infinite number of solutions)

Example

$x + 2y + 2z = 11$
$x - y + 3z = 8$
$4x - y + 11z = 35$

$$\begin{pmatrix} 1 & 2 & 2 & 11 \\ 1 & -1 & 3 & 8 \\ 4 & -1 & 11 & 35 \end{pmatrix} \rightarrow \begin{pmatrix} 1 & 2 & 2 & 11 \\ 0 & -3 & 1 & -3 \\ 0 & 0 & 0 & 0 \end{pmatrix}$$

so $-3y + z = -3$ and $x + 2y + 2z = 11$

$\Rightarrow y = \dfrac{z+3}{3}$ $x = 11 - 2y - 2z$

$\Rightarrow x = 11 - 2\left(\dfrac{z+3}{3}\right) - 2z$

$\Rightarrow x = 13 - \dfrac{8z}{3} = \dfrac{39 - 8z}{3}$

$z = z$

Parametric equations are found

- **Two lines of intersection** (an infinite number of solutions)

Such a system reduces to a form similar to the following:

$$\begin{pmatrix} 1 & 2 & 3 & 11 \\ 0 & 1 & 1 & -3 \\ 0 & 0 & 0 & 5 \end{pmatrix}$$

The system is inconsistent
Parallel planes are found by using the second row
to make parametric equations.

- **Three lines of intersection**

Similar to above.
Examine each pair of planes in turn.

- **A plane of intersection** (two redundant equations)

Such a system reduces to
a form similar to the following:

$$\begin{pmatrix} 1 & 2 & 3 & 11 \\ 0 & 0 & 0 & 0 \\ 0 & 0 & 0 & 0 \end{pmatrix}$$

No intersection

Such a system reduces to
a form similar to the following:

$$\begin{pmatrix} 1 & 2 & 3 & 11 \\ 0 & 0 & 0 & 8 \\ 0 & 0 & 0 & 5 \end{pmatrix}$$

No consistency
All planes are parallel

PARTIAL FRACTIONS

MAC 1.1: Applying algebraic skills to partial fractions
Part Subtitle

PARTIAL FRACTIONS

A proper rational function is one in which the degree of the numerator is less than that of the denominator.

eg. $\dfrac{x+3}{x^2}$

All others are termed improper.
These can be simplified through long division.

$\dfrac{x^2+3x+8}{x+5} =$ Simplify by dividing

$$= x+5 \overline{\smash{\big)}\begin{array}{r} x-2 \\ x^2+3x+8 \\ \underline{x^2+5x} \\ -2x+8 \\ \underline{-2x-10} \\ 18 \end{array}}$$

$\dfrac{x^2+3x+8}{x+5} = x-2+\dfrac{18}{x+5}$

General forms

Linear denominator	Quadratic denominator
$\dfrac{a}{bx+c}$	$\dfrac{ax+b}{cx^2+dx+e}$
$b \neq 0, \quad bx+c \neq 0$	$c \neq 0, \quad cx^2+dx+e \neq 0$

DISTINCT LINEAR FACTORS

Example

Express $\dfrac{5x-3}{x^2+x-30}$ in partial fractions

First, write the denominator in factorised form

$$\frac{5x-3}{x^2+x-30} = \frac{5x-3}{(x-5)(x+6)}$$

Now set this equal to two partial fractions

$$\frac{5x-3}{(x-5)(x+6)} = \frac{A}{x-5} + \frac{B}{x+6}$$

Thus

$$\frac{5x-3}{(x-5)(x+6)} \equiv \frac{A(x+6)+B(x-5)}{(x-5)(x+6)}$$

This is an identity, so the values of A and B can be found by comparing coefficients

$5x - 3 = A(x+6) + B(x-5)$

Pick values of x to find A & B

Let $x = -6$

$-33 = -11B$

$B = 3$

Let $x = 5$

$22 = 11A$

$A = 2$

$$\frac{5x-3}{(x-5)(x+6)} = \frac{2}{x-5} + \frac{3}{x+6}$$

REPEATED LINEAR FACTOR

Example

Express $\dfrac{3x^2 - 11x + 5}{(x-2)(x-1)^2}$ in partial fractions

$$\dfrac{3x^2 - 11x + 5}{(x-2)(x-1)^2} = \dfrac{A}{x-2} + \dfrac{B}{x-1} + \dfrac{C}{(x-1)^2}$$

$$= \dfrac{A(x-1)^2 + B(x-2)(x-1) + C(x-2)}{(x-2)(x-1)^2}$$

$$3x^2 - 11x + 5 \equiv A(x-1)^2 + B(x-2)(x-1) + C(x-2)$$

Comparing coefficients

Let $x = 1$
$-3 = -C$
$C = 3$

Let $x = 2$
$-5 = A$
$A = -5$

Let $x = 0$
$5 = A + 2B - 2C$
$5 = -5 + 2B - 6$
$16 = 2B$
$B = 8$

Substitute

$$\dfrac{3x^2 - 11x + 5}{(x-2)(x-1)^2} = \dfrac{-5}{x-2} + \dfrac{8}{x-1} + \dfrac{3}{(x-1)^2}$$

IRREDUCIBLE QUADRATIC FACTOR

Example

Express $\dfrac{4x^2 - 4x + 1}{(x-2)(x^2 - x + 1)}$ in partial fractions

$$\frac{4x^2 - 4x + 1}{(x-2)(x^2 - x + 1)} = \frac{A}{x-2} + \frac{Bx + C}{x^2 - x + 1}$$

$$= \frac{A(x^2 - x + 1) + (Bx + C)(x - 2)}{(x-2)(x^2 - x + 1)}$$

$$4x^2 - 4x + 1 = A(x^2 - x + 1) + (Bx + C)(x - 2)$$

comparing coefficients

Let $x = 2$
$9 = 3A$
$A = 3$

Let $x = 1$
$1 = A - (B + C)$
$B = A - 1 - C$
$B = 1$

Let $x = 0$
$1 = A - 2C$
$2C = 2$
$C = 1$

Substitute

$$\frac{4x^2 - 4x + 1}{(x-2)(x^2 - x + 1)} = \frac{3}{x-2} + \frac{x+1}{x^2 - x + 1}$$

CALCULUS

MAC 1.2: Applying calculus skills through techniques of differentiation.

MAC 1.3: Applying calculus skills through techniques of integration.

MAC 1.4: Applying calculus skills to solving differential equations.

DIFFERENTIATION

Differential Calculus concerns the rate of change of a function with respect to the change in the variable on which it depends.

For example,

Speed is a measure of distance travelled by an object in a unit period of time.

The distance travelled totally depends on the time taken.

Instantaneous speed is the speed of an object at a particular time.

Average speed = total distance divided by total time taken.

Velocity is a measure of the distance travelled by an object in a particular direction over a unit period of time.

Acceleration is a measure of how the velocity of an object changes with time.

The derivative of a function for some particular value is a measure of the rate at which the function is changing at that particular value.

The derivative of a function for some particular value is also the gradient of the graph of the function at that point.

FINDING GRADIENTS OF CURVES

The gradient of a straight line is found by dividing the change in the y-axis (the y-step) by the change in the x-axis (the x-step).

But what happens if we need to find the gradient of a curve?

The points A(2,4) and B(3,9) lie on the curve $y = x^2$.

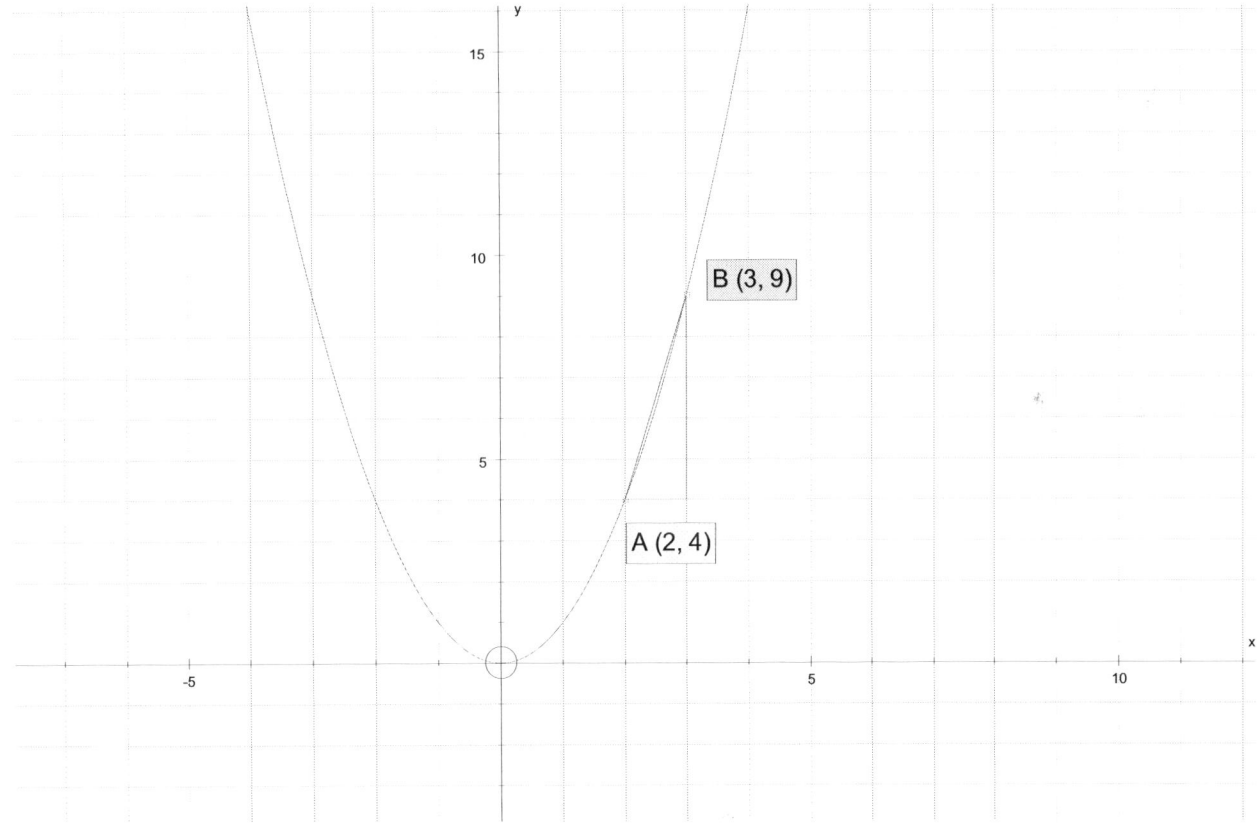

The average gradient of the curve from A to B is the gradient of the chord AB.

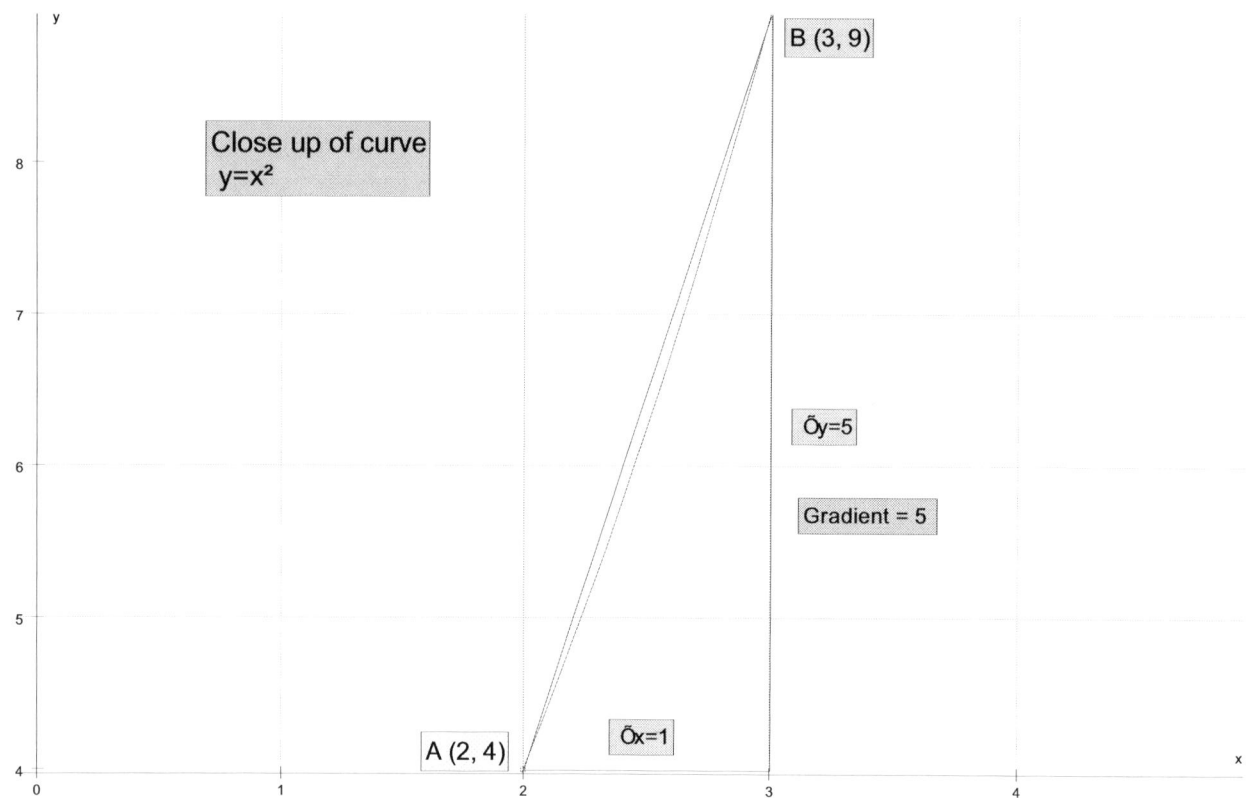

$$m_{AB} = \frac{y_2 - y_1}{x_2 - x_1}$$

$$m_{AB} = \frac{9-4}{3-2}$$

$$m_{AB} = 5$$

$\Delta y =$ change in y

$\Delta x =$ change in x

$$m_{AB} = \frac{y_2 - y_1}{x_2 - x_1} = \frac{\Delta y}{\Delta x}$$

As B moves towards A, the gradient of the curve changes.

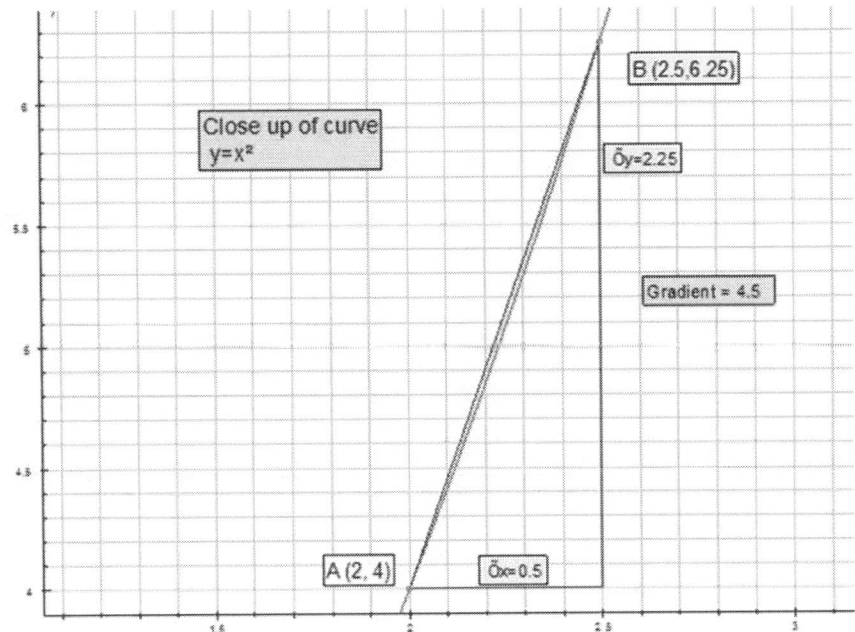

$$m_{AB} = \frac{y_2 - y_1}{x_2 - x_1}$$

$$m_{AB} = \frac{6.25 - 4}{2.5 - 2}$$

$$m_{AB} = \frac{2.25}{0.5}$$

$$m_{AB} = 4.5$$

$$\frac{\Delta y}{\Delta x} = 4.5$$

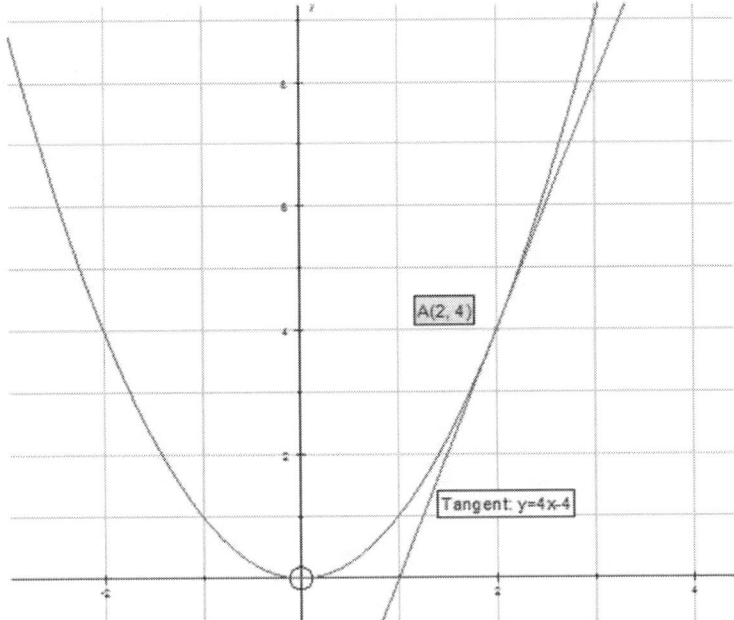

$$m_{AB} = \frac{y_2 - y_1}{x_2 - x_1}$$

$$m_{AB} = \frac{4.41 - 4}{2.1 - 2}$$

$$m_{AB} = \frac{0.41}{0.1}$$

$$m_{AB} = 4.1$$

$$\frac{\Delta y}{\Delta x} = 4.1$$

The closer B is to A, the nearer it reaches a limit

This limit is the gradient of the curve and occurs at the tangent to the curve at the point A.

x	f(x)	Δy	Δx	m = Δy/Δx
2.001	4.004001	0.004001	0.001	4.001
2.0009	4.003601	0.003601	0.0009	4.0009
2.0008	4.003201	0.003201	0.0008	4.0008
2.0007	4.0028	0.0028	0.0007	4.0007
2.0006	4.0024	0.0024	0.0006	4.0006
2.0005	4.002	0.002	0.0005	4.0005
2.0004	4.0016	0.0016	0.0004	4.0004
2.0003	4.0012	0.0012	0.0003	4.0003
2.0002	4.0008	0.0008	0.0002	4.0002
2.0001	4.0004	0.0004	1E-04	4.0001

When A is the point (2, 4), the gradient of the curve is 4.

Other points have different gradients.

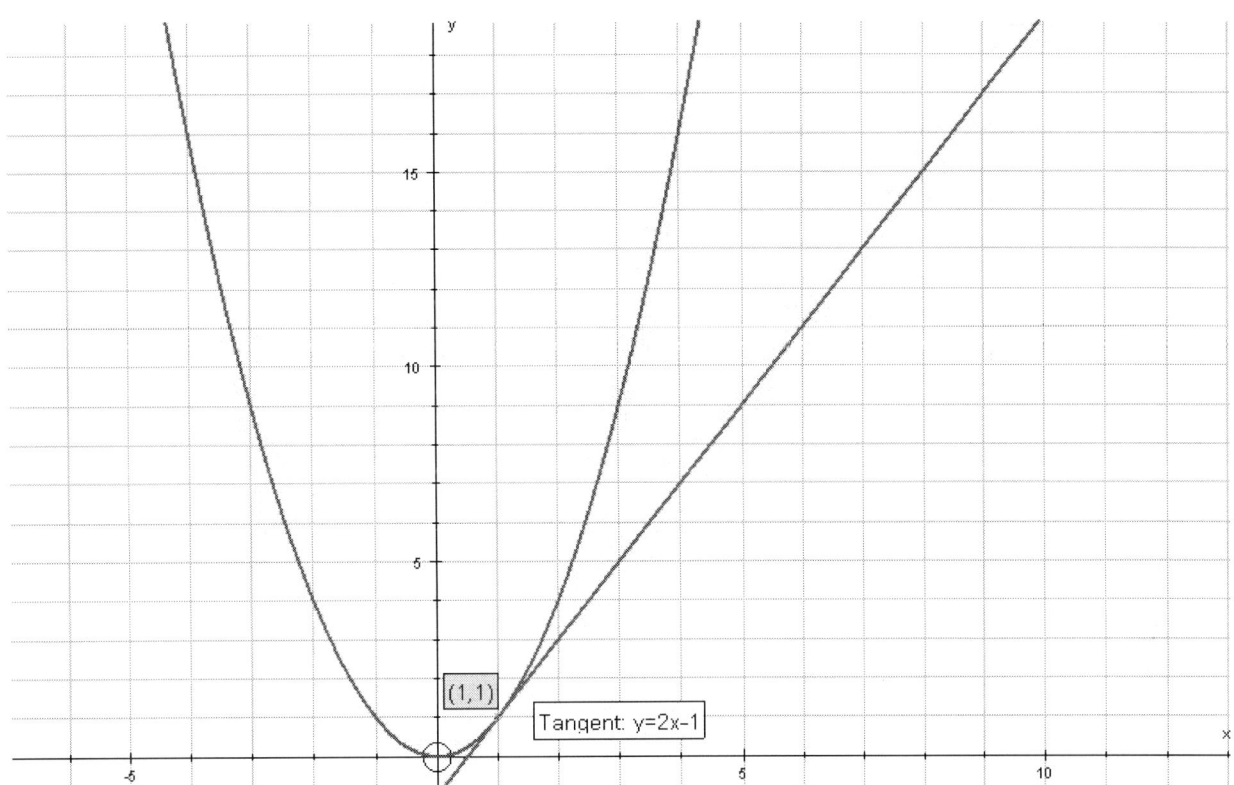

Is there an easier way to find the gradient of the curve at other points?

Let A(x, f(x)) and B(x+h,(f(x+h)) be points on the graph y = f(x)

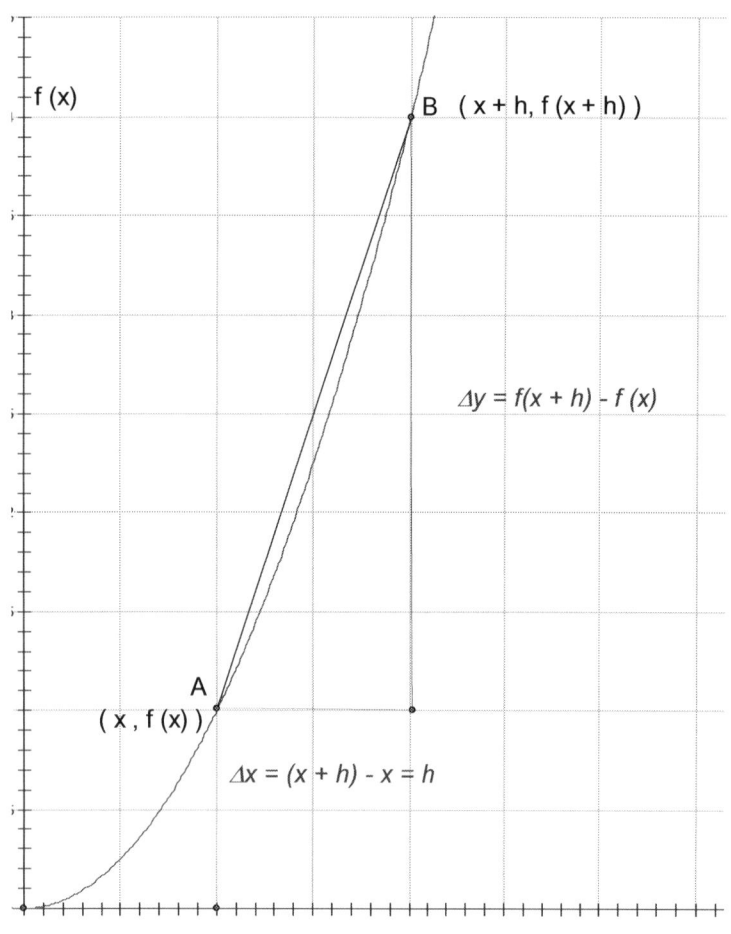

$$m_{AB} = \frac{y_2 - y_1}{x_2 - x_1}$$

$$m_{AB} = \frac{f(x+h) - f(x)}{(x+h) - x}$$

$$m_{AB} = \frac{f(x+h) - f(x)}{h}$$

$$m_{AB} = \frac{y_2 - y_1}{x_2 - x_1}$$

$$m_{AB} = \frac{\Delta y}{\Delta x} = \lim_{\Delta x \to 0} \frac{\Delta y}{\Delta x}$$

$$m_{AB} = \frac{dy}{dx} \quad \text{Leibniz notation}$$

The gradient is derived by reducing the difference between x and h.

As h tends to zero, the gradient tends to a limit.

This limit is the gradient of the tangent to the curve at the point A.

$$f'(x) = \lim_{h \to 0} \frac{f(x+h) - f(x)}{h}$$

$f'(x)$ is called the derivative or derived function of f(x)

This is the rate of change of the function and the gradient of the tangent to the graph of the function

Leibniz notation can also be used.

Instead of writing $f'(x)$, write $\dfrac{dy}{dx}$

$$\frac{dy}{dx} = \lim_{h \to 0} \frac{f(x+h) - f(x)}{h}$$

DIFFERENTIATION BY FIRST PRINCIPLES – DERIVING F'(X) FROM F(X)

$$f'(x) = \lim_{h \to 0} \frac{f(x+h) - f(x)}{h}$$

Example
If $f(x) = x^2$, find $f'(x)$

$$f'(x) = \lim_{h \to 0} \frac{f(x+h) - f(x)}{h}$$
$$= \lim_{h \to 0} \frac{(x+h)^2 - x^2}{h}$$
$$= \lim_{h \to 0} \frac{x^2 + 2hx + h^2 - x^2}{h}$$
$$= \lim_{h \to 0} \frac{2hx + h^2}{h}$$
$$= \lim_{h \to 0} \frac{h(2x+h)}{h}$$
$$= \lim_{h \to 0} 2x + h$$
$$= 2x$$

The derived function of x is $f'(x) = 2x$
The derivative when $x = 3$ is 6

This means that the rate of change of the function is 2x, and that the gradient of the tangent to the graph of the curve $y = x^2$ at any point is found by doubling the x co-ordinate.

$f(-1) = 1$ $f'(-1) = -2$ has tangent $y - 1 = -2(x-(-1))$ i.e. $y = -2x - 1$
$f(0) = 0$ $f'(0) = 0$ has tangent $y - 0 = 0(x-0)$ i.e. $y = 0$

f(1) = 1   f'(1) = 2   has tangent y−1 = 2(x−1) i.e. y = 2x−1
f(2) = 4   f'(2) = 4   has tangent y−4 = 4(x−2) i.e. y = 4x−4
f(3) = 9   f'(3) = 6   has tangent y−9 = 6(x−3) i.e. y = 6x−9
f(4) = 16   f'(4) = 8   has tangent y−16 = 8(x−4) i.e. y = 8x−16

The gradient of the tangent to the graph at the point (3, 9) is 6.

The gradient of the tangent to the graph at the point (4, 16) is 8.

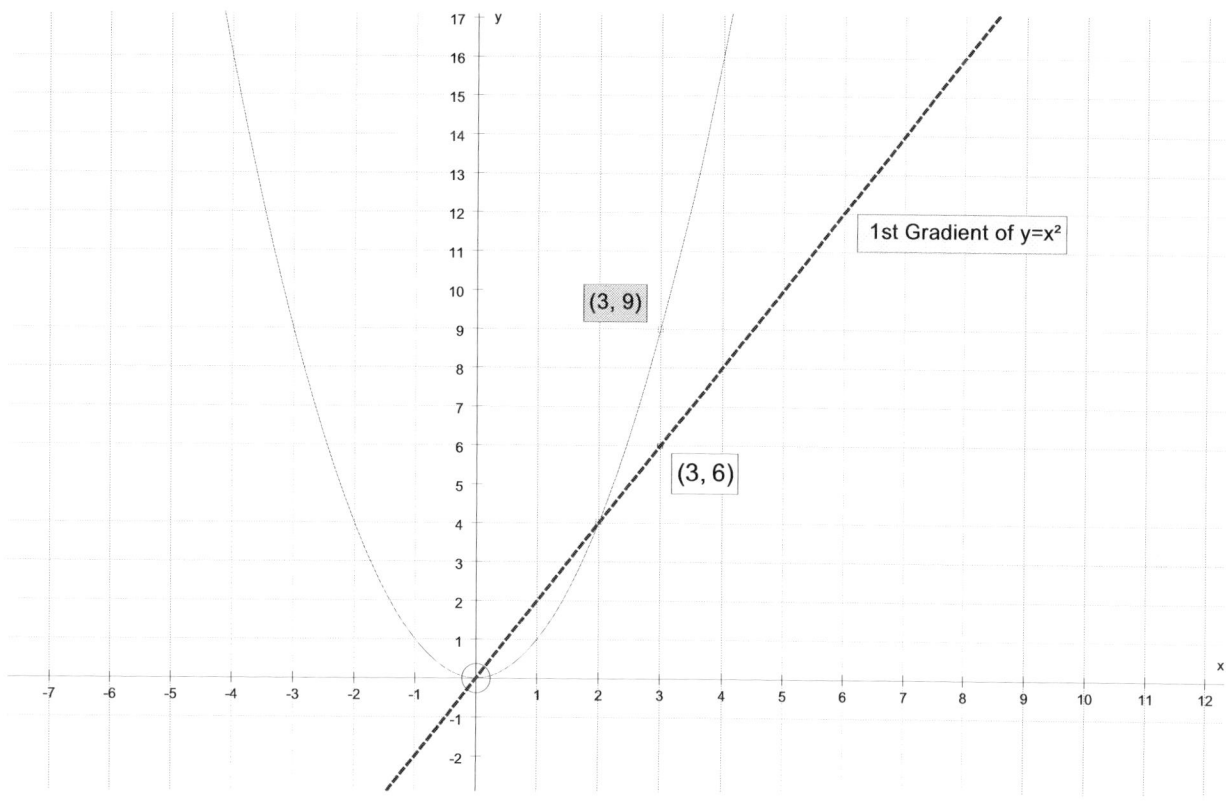

In General

$$\boxed{\text{If } f(x) = x^n \text{ then } f'(x) = nx^{n-1}\\ n \text{ is a rational number}}$$

Example

If $f(x) = x^3$, find $f'(x)$

$$f'(x) = \lim_{h \to 0} \frac{f(x+h) - f(x)}{h}$$

$$= \lim_{h \to 0} \frac{(x+h)^3 - x^3}{h}$$

$$= \lim_{h \to 0} \frac{x^3 + 3hx^2 + 3h^2 x + h^3 - x^3}{h}$$

$$= \lim_{h \to 0} \frac{3hx^2 + 3h^2 x + h^3}{h}$$

$$= \lim_{h \to 0} \frac{h\left(3x^2 + 3h\,x + h^2\right)}{h}$$

$$= \lim_{h \to 0} 3x^2 + 3h\,x + h^2$$

$$= 3x^2$$

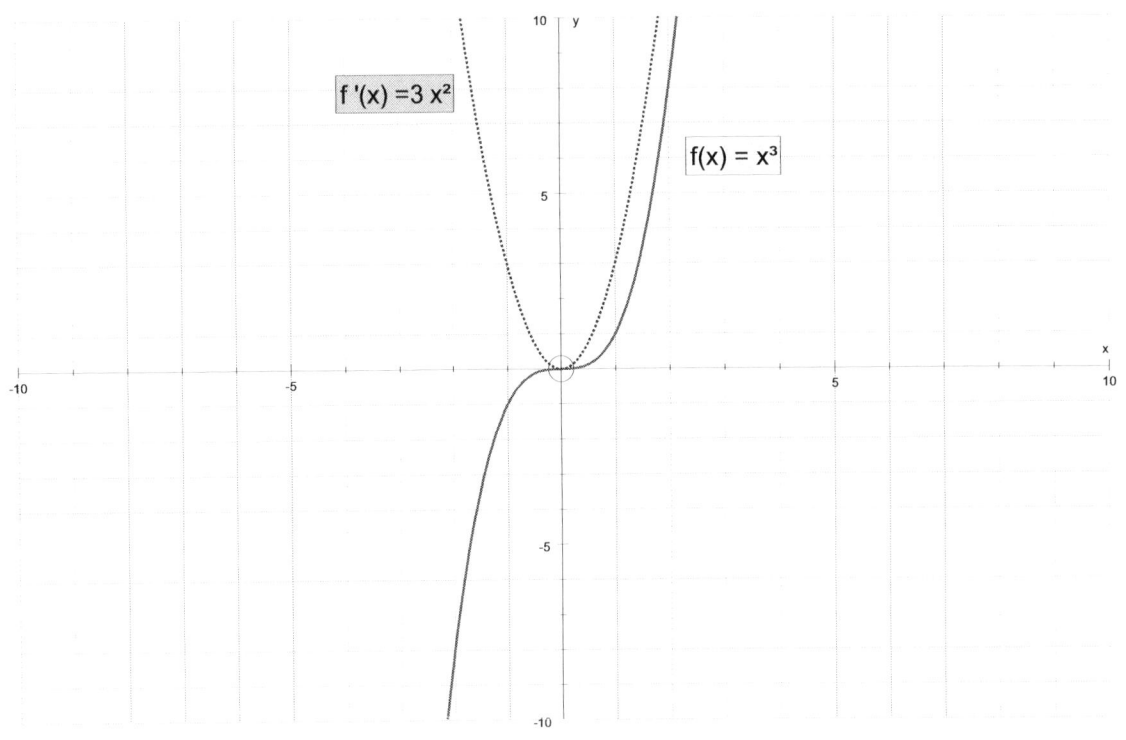

FINDING OTHER DERIVATIVES BY FIRST PRINCIPLES

$$\text{If } f(x) = g(x) + h(x), f'(x) = g'(x) + h'(x)$$

Examples

If $f(x) = x^2 + 2$, find $f'(x)$

$$f'(x) = \lim_{h \to 0} \frac{f(x+h) - f(x)}{h}$$
$$= \lim_{h \to 0} \frac{((x+h)^2 + 2) - (x^2 + 2)}{h}$$
$$= \lim_{h \to 0} \frac{x^2 + 2hx + h^2 + 2 - x^2 - 2}{h}$$
$$= \lim_{h \to 0} \frac{2hx + h^2}{h}$$
$$= \lim_{h \to 0} \frac{h(2x + h)}{h}$$
$$= \lim_{h \to 0} 2x + h$$
$$= 2x$$

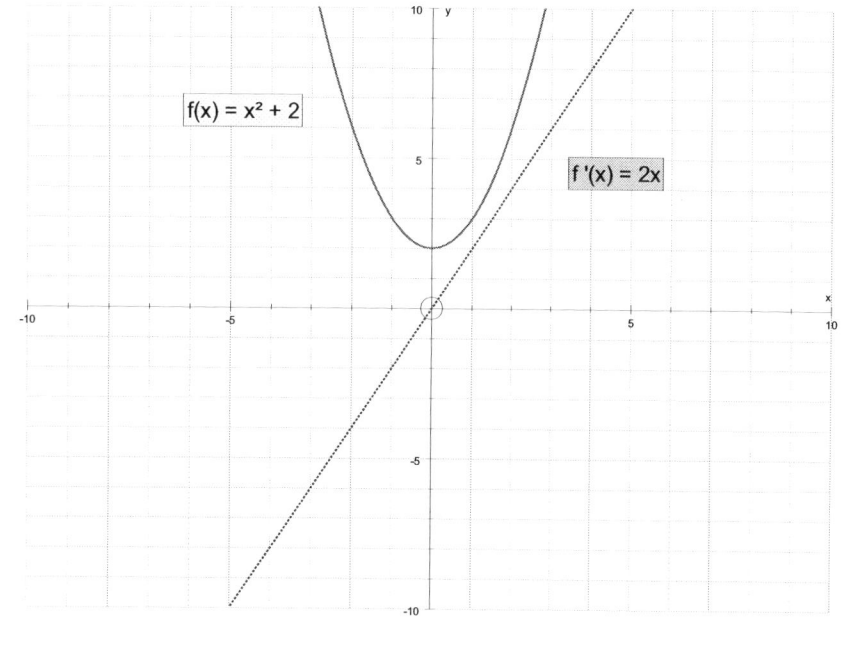

If $f(x) = x^2 + 2x$, find $f'(x)$

$$f'(x) = \lim_{h \to 0} \frac{f(x+h) - f(x)}{h}$$

$$= \lim_{h \to 0} \frac{\left((x+h)^2 + 2(x+h)\right) - \left(x^2 + 2x\right)}{h}$$

$$= \lim_{h \to 0} \frac{x^2 + 2hx + h^2 + 2x + 2h - x^2 - 2x}{h}$$

$$= \lim_{h \to 0} \frac{2hx + h^2 + 2h}{h}$$

$$= \lim_{h \to 0} \frac{h(2x + h + 2)}{h}$$

$$= \lim_{h \to 0} 2x + h + 2$$

$$= 2x + 2$$

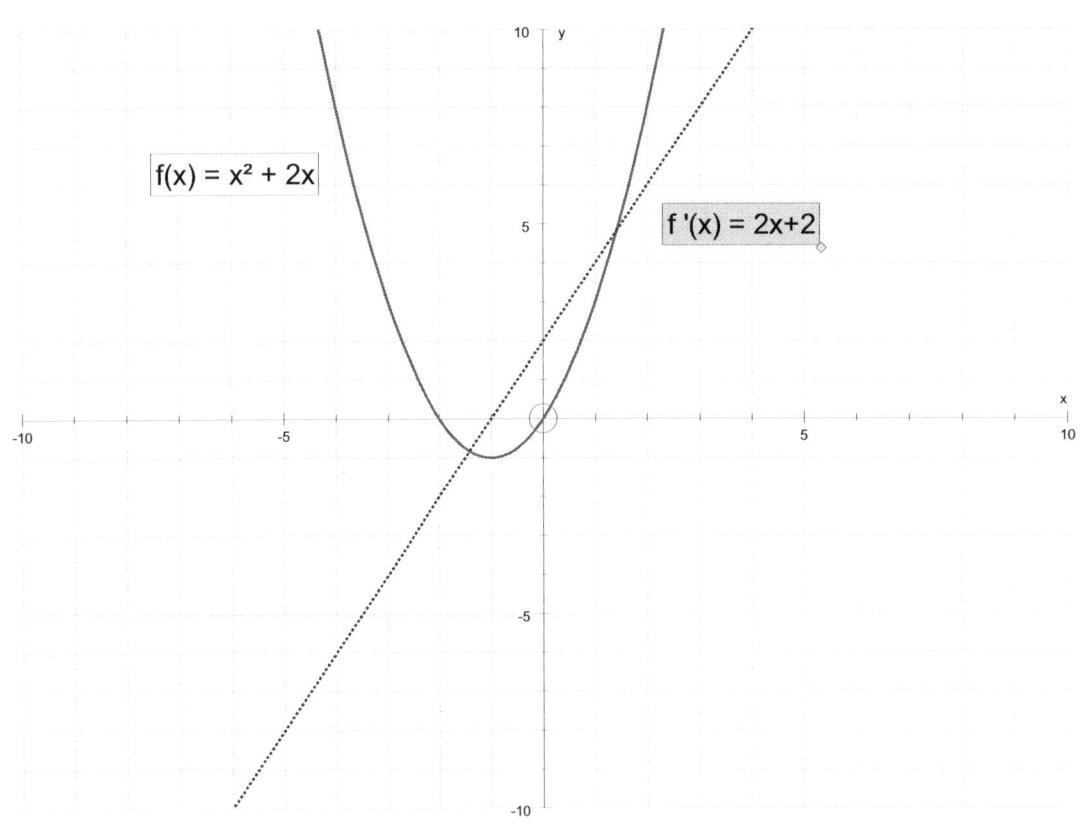

If f(x) = x² +2x +3, find f' (x)

$$f'(x) = \lim_{h \to 0} \frac{f(x+h) - f(x)}{h}$$

$$= \lim_{h \to 0} \frac{\left((x+h)^2 + 2(x+h) + 3\right) - \left(x^2 + 2x + 3\right)}{h}$$

$$= \lim_{h \to 0} \frac{x^2 + 2hx + h^2 + 2x + 2h + 3 - x^2 - 2x - 3}{h}$$

$$= \lim_{h \to 0} \frac{2hx + h^2 + 2h}{h}$$

$$= \lim_{h \to 0} \frac{h(2x + h + 2)}{h}$$

$$= \lim_{h \to 0} 2x + h + 2$$

$$= 2x + 2$$

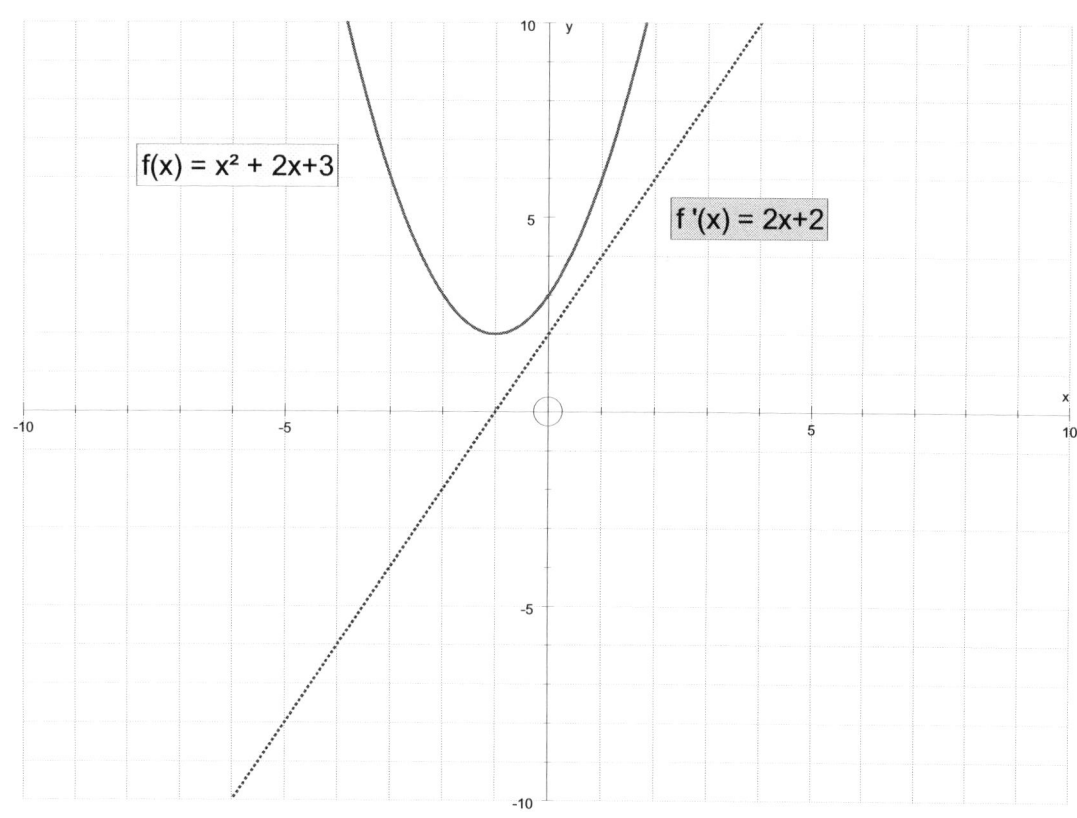

If f(x) = x³ + 2x² − 3x + 4, find f'(x)

$$f'(x) = \lim_{h \to 0} \frac{f(x+h) - f(x)}{h}$$

$$= \lim_{h \to 0} \frac{\left((x+h)^3 + 2(x+h)^2 - 3(x+h) + 4\right) - \left(x^3 + 2x^2 - 3x + 4\right)}{h}$$

$$= \lim_{h \to 0} \frac{(x^3 + 3hx^2 + 3h^2x + h^3) + 2(x^2 + 2xh + h^2) - 3(x+h) + 4 - \left(x^3 + 2x^2 - 3x + 4\right)}{h}$$

$$= \lim_{h \to 0} \frac{x^3 + 3hx^2 + 3h^2x + h^3 + 2x^2 + 4xh + 2h^2 - 3x - 3h + 4 - x^3 - 2x^2 + 3x - 4}{h}$$

$$= \lim_{h \to 0} \frac{\cancel{x^3} + 3hx^2 + 3h^2x + h^3 \cancel{+ 2x^2} + 4xh + 2h^2 \cancel{- 3x} - 3h \cancel{+ 4} \cancel{- x^3} \cancel{- 2x^2} \cancel{+ 3x} \cancel{- 4}}{h}$$

$$= \lim_{h \to 0} \frac{3hx^2 + 3h^2x + h^3 + 4xh + 2h^2 - 3h}{h}$$

$$= \lim_{h \to 0} \frac{h(3x^2 + 3hx + h^2 + 4x + 2h - 3)}{h}$$

$$= \lim_{h \to 0} 3x^2 + 3hx + h^2 + 4x + 2h - 3$$

$$= 3x^2 + 4x - 3$$

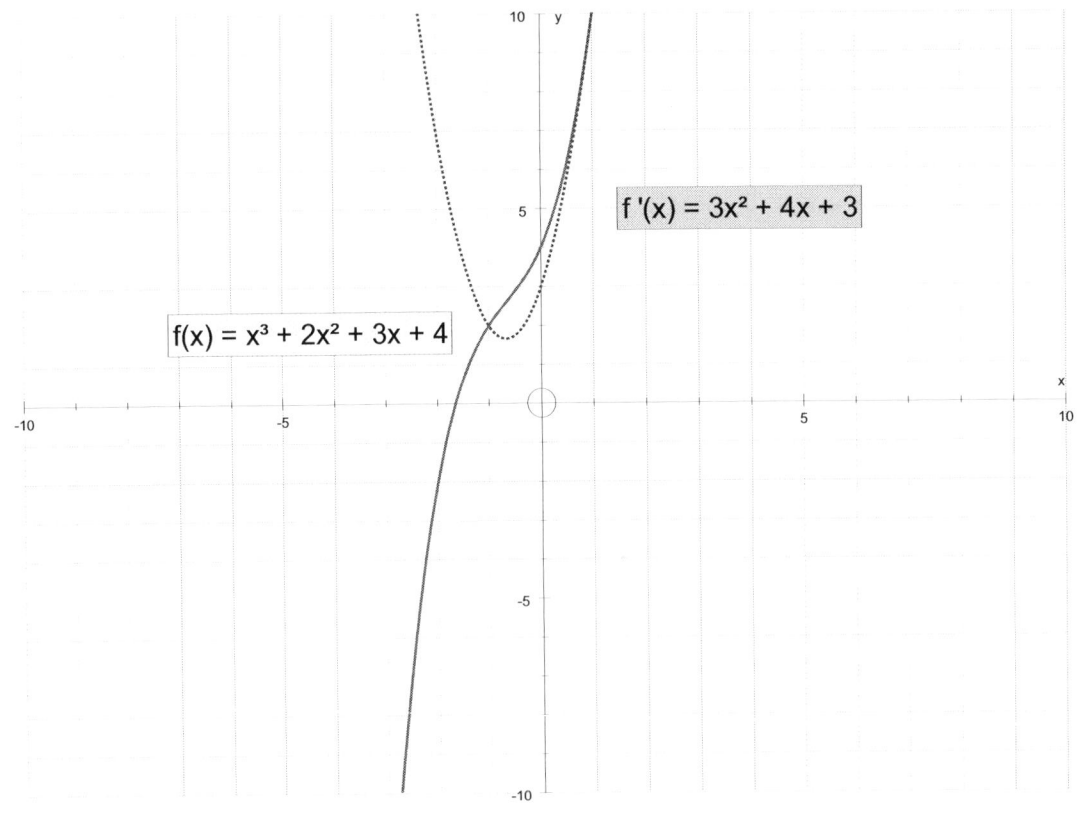

$$\boxed{\text{If } f(x) = ax^n \text{ then } f'(x) = anx^{n-1}\\ a \text{ is a constant}, \quad n \text{ is a rational number}}$$

Example

If $f(x) = 3x^2$, find $f'(x)$

$$f'(x) = \lim_{h \to 0} \frac{f(x+h) - f(x)}{h}$$
$$= \lim_{h \to 0} \frac{(3(x+h)^2) - (3x^2)}{h}$$
$$= \lim_{h \to 0} \frac{3(x^2 + 2hx + h^2) - 3x^2}{h}$$
$$= \lim_{h \to 0} \frac{6hx + 3h^2}{h}$$
$$= \lim_{h \to 0} \frac{h(6x + 3h)}{h}$$
$$= \lim_{h \to 0} 6x + 3h$$
$$= 6x$$

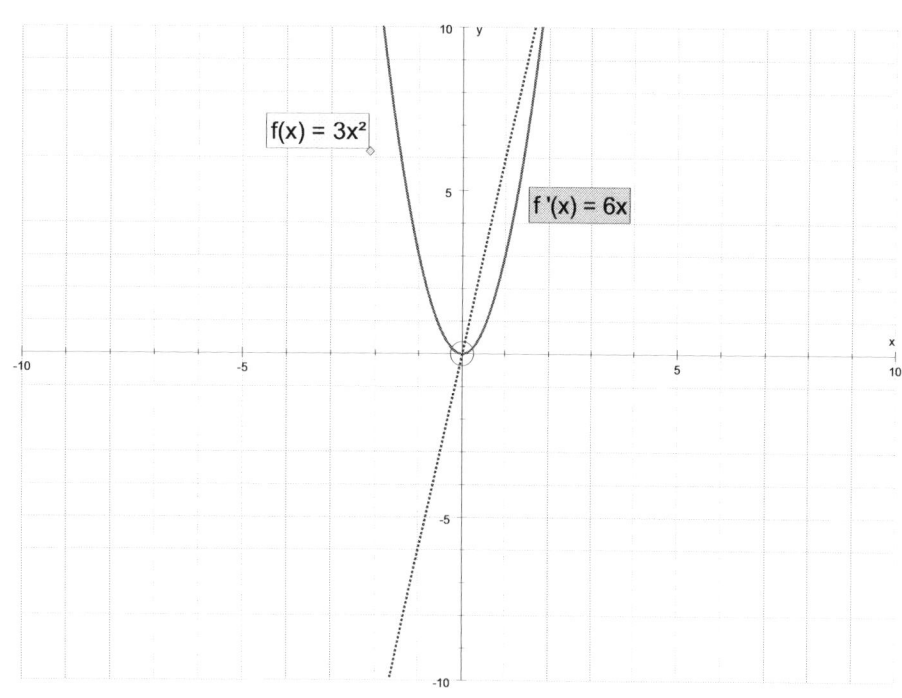

> If $f(x) = (x+a)^n$ then $f'(x) = n(x+a)^{n-1}$
> a is a real number, n is a rational number

Example
If $f(x) = (x+3)^2$, find $f'(x)$

$$f'(x) = \lim_{h\to 0}\frac{f(x+h)-f(x)}{h}$$
$$= \lim_{h\to 0}\frac{((x+h)+3)^2 - (x+3)^2}{h}$$
$$= \lim_{h\to 0}\frac{(x+h+3)^2 - (x+3)^2}{h}$$
$$= \lim_{h\to 0}\frac{x^2 + hx + 3x + hx + h^2 + 3h + 3x + 3h + 9 - (x+3)^2}{h}$$
$$= \lim_{h\to 0}\frac{x^2 + 2hx + 6x + 6h + h^2 + 9 - [x^2 + 6x + 9]}{h}$$
$$= \lim_{h\to 0}\frac{x^2 + 2hx + 6x + 6h + h^2 + 9 - x^2 - 6x - 9}{h}$$
$$= \lim_{h\to 0}\frac{2hx + 6h + h^2}{h}$$
$$= \lim_{h\to 0}\frac{h(2x+6+h)}{h}$$
$$= \lim_{h\to 0} 2x + 6 + h$$
$$= 2x + 6$$
$$= 2(x+3)$$

f(x) = (x + 3)²   f '(x) = 2(x + 3)

> If $f(x) = (ax+b)^n$ then $f'(x) = an(ax+b)^{n-1}$
> a is a real number, n is a rational number

THE CHAIN RULE

$$h'(x) = g'(f(x)) \times f'(x)$$

Let $h(x) = (ax+b)^n = g(f(x))$
where $f(x) = ax+b$ and $g(x) = x^n$
Then $f'(x) = a$ and $g'(x) = nx^{n-1}$
$g'(ax+b) = n(ax+b)^{n-1}$
$g'(f(x)) = n(ax+b)^{n-1}$

$h(x) = (ax+b)^n$
$h'(x) = an(ax+b)^{n-1}$
$\Rightarrow h'(x) = a \times n(ax+b)^{n-1}$
$= f'(x) \times n(ax+b)^{n-1}$
$= f'(x) \times g'(f(x))$

In Leibniz notation

$y = (ax+b)^n$
Let $u = ax+b$ then $y = u^n$

$\dfrac{du}{dx} = a$ and $\dfrac{dy}{du} = nu^{n-1} = n(ax+b)^{n-1}$

$\dfrac{dy}{dx} = an(ax+b)^{n-1}$
$= a \times n(ax+b)^{n-1}$
$= \dfrac{du}{dx} \times \dfrac{dy}{du}$

$$\dfrac{dy}{dx} = \dfrac{du}{dx} \times \dfrac{dy}{du}$$

PRODUCT RULE : FIRST PRINCIPLES

$$\text{If } f(x) = g(x).h(x) \text{ then } f'(x) = g'(x).h(x) + g(x).h'(x)$$

Example

If $f(x) = 2x(5x+3)^2$, find $f'(x)$

$$f'(x) = \lim_{h \to 0} \frac{f(x+h) - f(x)}{h}$$
$$= \lim_{h \to 0} \frac{2(x+h)(5(x+h)+3)^2 - 2x(5x+3)^2}{h}$$
$$= \lim_{h \to 0} \frac{(2x+2h)(5x+5h+3)^2 - 2x(5x+3)^2}{h}$$
$$= \lim_{h \to 0} \frac{(2x+2h)(25x^2 + 25hx + 15x + 25hx + 25h^2 + 15h + 15x + 15h + 9) - 2x(25x^2 + 30x + 9)}{h}$$
$$= \lim_{h \to 0} \frac{(2x+2h)(25x^2 + 50hx + 30x + 25h^2 + 30h + 9) - (50x^3 + 60x^2 + 18x)}{h}$$
$$= \lim_{h \to 0} \frac{(50x^3 + 100hx^2 + 60x^2 + 50h^2x + 60hx + 18x + 50x^2h + 100h^2x + 60hx + 50h^3 + 60h^2 + 18h) - (50x^3 + 60x^2 + 18x)}{h}$$
$$= \lim_{h \to 0} \frac{50x^3 + 150hx^2 + 60x^2 + 120hx + 18x + 150h^2x + 50h^3 + 60h^2 + 18h - 50x^3 - 60x^2 - 18x}{h}$$
$$= \lim_{h \to 0} \frac{150hx^2 + 120hx + 150h^2x + 50h^3 + 60h^2 + 18h}{h}$$
$$= \lim_{h \to 0} \frac{h(150x^2 + 120x + 150hx + 50h^2 + 60h + 18)}{h}$$
$$= \lim_{h \to 0} 150x^2 + 120x + 150hx + 50h^2 + 60h + 18$$
$$= 150x^2 + 120x + 18$$
$$= 50x^2 + 60x + 18 + 100x^2 + 60x$$
$$= 2(25x^2 + 30x + 9) + 100x^2 + 60x$$
$$= 2(5x+3)^2 + 20x(5x+3)$$

QUOTIENT RULE

$$\text{If} \quad f(x) = \frac{g(x)}{h(x)}$$

$$\text{then}, \quad f'(x) = \left(\frac{g(x)}{h(x)}\right)' = \frac{g'(x)h(x) - g(x)h'(x)}{(g(x))^2}$$

This is derived from the product rule

Example

$f(x) = \dfrac{x^4}{x+1}$, find $f'(x)$

$f(x) = \dfrac{x^4}{(x+1)} = x^4(x+1)^{-1}$

Applying product rule

$f'(x) = 4x^3(x+1)^{-1} + x^4 \cdot -1(x+1)^{-2}$

$\quad = 4x^3(x+1)^{-1} - x^4(x+1)^{-2}$

$\quad = \dfrac{4x^3}{(x+1)} - \dfrac{x^4}{(x+1)^2}$

$\quad = \dfrac{4x^3(x+1) - x^4}{(x+1)^2}$

$\quad = \dfrac{4x^4 + 4x^3 - x^4}{(x+1)^2}$

$\quad = \dfrac{3x^4 + 4x^3}{(x+1)^2}$

$\quad = \dfrac{x^3(3x+4)}{(x+1)^2}$

By just using the quotient rule

$f(x) = \dfrac{g(x)}{h(x)}$ where $g(x) = x^4$ and $h(x) = x+1$

$f'(x) = \left(\dfrac{g(x)}{h(x)}\right)' = \dfrac{g'(x)h(x) - g(x)h'(x)}{(h(x))^2}$

$f'(x) = \dfrac{4x^3(x+1) - x^4}{(x+1)^2}$

$\quad = \dfrac{4x^4 + 4x^3 - x^4}{(x+1)^2}$

$\quad = \dfrac{x^3(3x+4)}{(x+1)^2}$

FINDING TRIGONOMETRIC DERIVATIVES BY FIRST PRINCIPLES

Using Radians

$$\lim_{h \to 0} \frac{\sin(h)}{h} = 1 \qquad \lim_{h \to 0} \frac{\cos(h)-1}{h} = 0$$

$$\text{If } f(x) = \sin x \text{ then } f'(x) = \cos x$$

Proof
If $f(x) = \sin x$, find $f'(x)$

$$f'(x) = \lim_{h \to 0} \frac{f(x+h) - f(x)}{h}$$

$$= \lim_{h \to 0} \frac{\sin(x+h) - \sin x}{h}$$

$$= \lim_{h \to 0} \frac{\sin x \cosh + \cos x \sinh - \sin x}{h}$$

$$= \lim_{h \to 0} \frac{\sin x (\cosh - 1) + \cos x \sinh}{h}$$

$$= \lim_{h \to 0} \frac{\sin x (\cosh - 1)}{h} + \lim_{h \to 0} \frac{\cos x \sinh}{h}$$

$$= \sin x \lim_{h \to 0} \frac{(\cosh - 1)}{h} + \cos x \lim_{h \to 0} \frac{\sinh}{h}$$

$$= \sin x \times 0 + \cos x \times 1$$

since $\lim_{h \to 0} \frac{\cos(h) - 1}{h} = 0$ and $\lim_{h \to 0} \frac{\sin(h)}{h} = 1$

$$= \cos x$$

Alternatively

$$f'(x) = \lim_{h \to 0} \frac{f(x+h) - f(x)}{h}$$

$$= \lim_{h \to 0} \frac{\sin(x+h) - \sin x}{h}$$

Now use the trig formula

$$SinA - SinB = 2\cos\left(\frac{A+B}{2}\right)\sin\left(\frac{A-B}{2}\right)$$

with $A = x + h$ and $B = x$

$$= \lim_{h \to 0} \frac{2\cos\left(\frac{x+h+x}{2}\right)\sin\left(\frac{x+h-x}{2}\right)}{h}$$

$$= \lim_{h \to 0} \frac{2\cos\left(\frac{2x+h}{2}\right)\sin\left(\frac{h}{2}\right)}{h}$$

$$= \lim_{h \to 0} \frac{2\cos\left(x+\frac{h}{2}\right)\sin\left(\frac{h}{2}\right)}{h}$$

$$= \lim_{h \to 0} 2\cos\left(x+\frac{h}{2}\right) \times \frac{\sin\left(\frac{h}{2}\right)}{h}$$

$$= \lim_{h \to 0} \cos\left(x+\frac{h}{2}\right) \times \frac{\sin\left(\frac{h}{2}\right)}{\frac{h}{2}}$$

$$= \lim_{h \to 0} \cos\left(x+\frac{h}{2}\right) \times 1 \qquad \text{since } \lim_{h \to 0} \frac{\sin\left(\frac{h}{2}\right)}{\frac{h}{2}} = 1$$

$$= \cos x$$

Example

If f(x) = sin5x , find f' (x)

$$f'(x) = \lim_{h \to 0} \frac{f(x+h) - f(x)}{h}$$

$$= \lim_{h \to 0} \frac{\sin 5(x+h) - \sin 5x}{h}$$

$$= \lim_{h \to 0} \frac{\sin 5x \cos 5h + \cos 5x \sin 5h - \sin 5x}{h}$$

$$= \lim_{h \to 0} \frac{\sin 5x (\cos 5h - 1) + \cos 5x \sin 5h}{h}$$

$$= \lim_{h \to 0} \frac{\sin 5x (\cos 5h - 1)}{h} + \lim_{h \to 0} \frac{\cos 5x \sin 5h}{h}$$

$$= \sin 5x \lim_{h \to 0} \frac{(\cos 5h - 1)}{h} + \cos 5x \lim_{h \to 0} \frac{\sin 5h}{h}$$

Now multiply each term by 1 in the form $\frac{5}{5}$

$$= \sin 5x \lim_{h \to 0} \frac{5(\cos 5h - 1)}{5h} + \cos 5x \lim_{h \to 0} \frac{5 \sin 5h}{5h}$$

$$= \sin 5x \times 5 \lim_{h \to 0} \frac{(\cos 5h - 1)}{5h} + \cos 5x \times 5 \lim_{h \to 0} \frac{\sin 5h}{5h}$$

$$= \sin 5x \times 0 + \cos 5x \times 5$$

since $\lim_{h \to 0} \frac{\cos(5h) - 1}{5h} = 0$ and $\lim_{h \to 0} \frac{\sin(5h)}{5h} = 1$

$$= 5 \cos 5x$$

$$\boxed{\text{If } f(x) = \cos x \text{ then } f'(x) = -\sin x}$$

$$f'(x) = \lim_{h \to 0} \frac{f(x+h) - f(x)}{h}$$

$$= \lim_{h \to 0} \frac{\cos(x+h) - \cos x}{h}$$

$$= \lim_{h \to 0} \frac{\cos x \cos h - \sin x \sin h - \cos x}{h}$$

$$= \lim_{h \to 0} \frac{\cos x (\cos h - 1) - \sin x \sin h}{h}$$

$$= \lim_{h \to 0} \frac{\cos x (\cos h - 1)}{h} - \lim_{h \to 0} \frac{\sin x \sin h}{h}$$

$$= \cos x \lim_{h \to 0} \frac{(\cos h - 1)}{h} - \sin x \lim_{h \to 0} \frac{\sin h}{h}$$

$$= \cos x \times 0 - \sin x \times 1$$

since $\lim_{h \to 0} \frac{\sin(h)}{h} = 1$ and $\lim_{h \to 0} \frac{\cos(h) - 1}{h} = 0$

$$= -\sin x$$

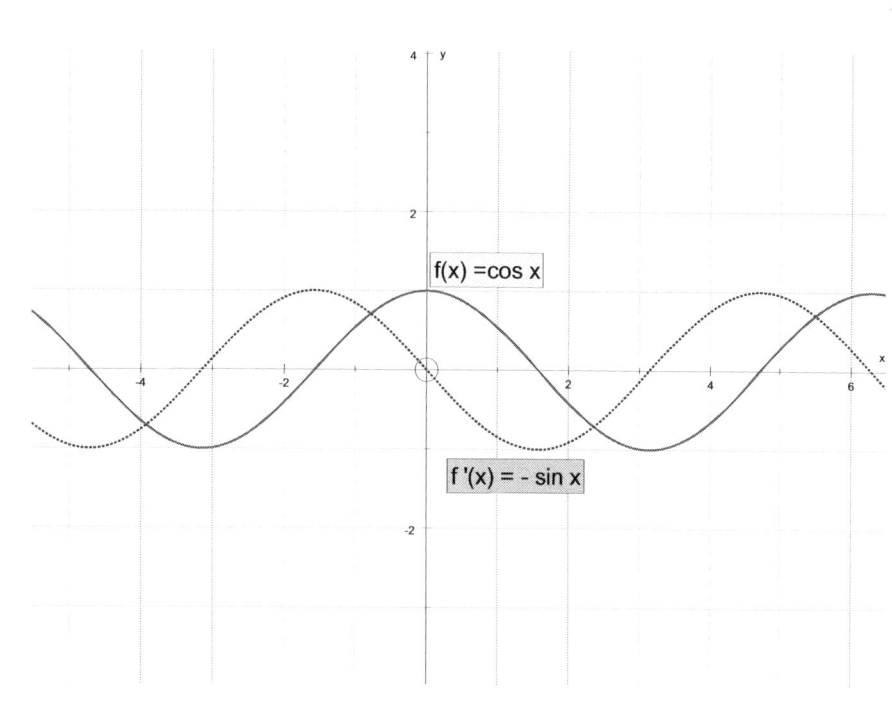

Alternatively

$$f'(x) = \lim_{h \to 0} \frac{f(x+h) - f(x)}{h}$$

$$= \lim_{h \to 0} \frac{\cos(x+h) - \cos x}{h}$$

Now use the trig formula

$$SinA - SinB = 2\cos\left(\frac{A+B}{2}\right)\sin\left(\frac{A-B}{2}\right)$$

with $A = x+h$ and $B = x$

$$= \lim_{h \to 0} \frac{-2\sin\left(\frac{x+h+x}{2}\right)\sin\left(\frac{x+h-x}{2}\right)}{h}$$

$$= \lim_{h \to 0} \frac{-2\sin\left(\frac{2x+h}{2}\right)\sin\left(\frac{h}{2}\right)}{h}$$

$$= \lim_{h \to 0} \frac{-2\sin\left(x + \frac{h}{2}\right)\sin\left(\frac{h}{2}\right)}{h}$$

$$= \lim_{h \to 0} -2\sin\left(x + \frac{h}{2}\right) \times \frac{\sin\left(\frac{h}{2}\right)}{h}$$

$$= \lim_{h \to 0} -\sin\left(x + \frac{h}{2}\right) \times \frac{\sin\left(\frac{h}{2}\right)}{\frac{h}{2}}$$

$$= \lim_{h \to 0} -\sin\left(x + \frac{h}{2}\right) \times 1 \qquad \text{since } \lim_{h \to 0} \frac{\sin\left(\frac{h}{2}\right)}{\frac{h}{2}} = 1$$

$$= -\sin x$$

$$\boxed{\text{If } f(x) = \tan x \text{ then } f'(x) = 1/\cos^2 x}$$

$$f(x) = \tan x = \frac{\sin x}{\cos x}$$

$$f'(x) = \frac{\cos x \cdot \cos x - \sin x \cdot -\sin x}{(\cos x)^2}$$

$$= \frac{\cos^2 x + \sin^2 x}{(\cos x)^2}$$

$$= \frac{1}{\cos^2 x}$$

$$= \sec^2 x$$

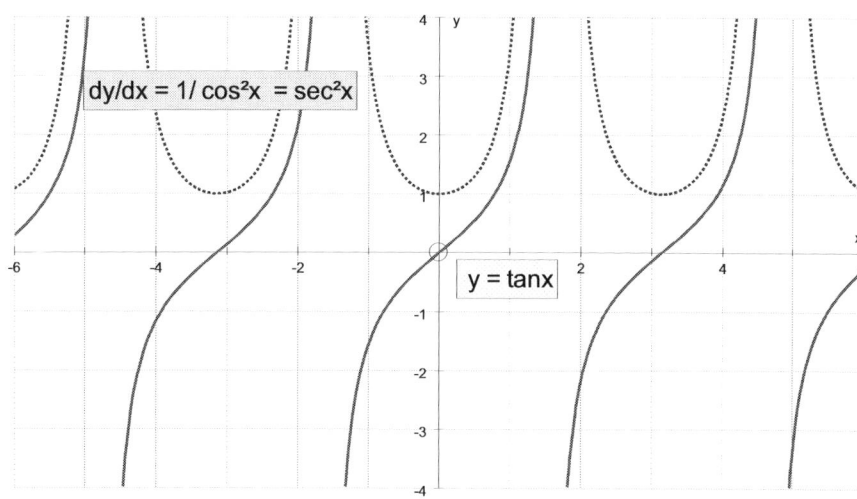

GRADIENTS OF TANGENTS TO CURVES

The derivative of a function for some particular value is also the gradient of the graph of the function at that point.

Example

A curve has equation $y = x - \dfrac{16}{\sqrt{x}}$, $x > 0$

Find the gradient of the tangent at the point where $x = 4$.

$y = x - \dfrac{16}{\sqrt{x}}$

$y = x - 16x^{-\frac{1}{2}}$

$\dfrac{dy}{dx} = 1 - \left(-\dfrac{1}{2} \times 16x^{-\frac{3}{2}}\right)$

$= 1 + 8x^{-\frac{3}{2}}$

$\dfrac{dy}{dx} = 1 + \dfrac{8}{\sqrt{x^3}}$

so when $x = 4$

$\dfrac{dy}{dx} = 1 + \dfrac{8}{\sqrt{x^3}}$

$= 1 + \dfrac{8}{\sqrt{4^3}}$

$= 1 + \dfrac{8}{\sqrt{64}}$

$= 1 + \dfrac{8}{8}$

$= 2$

Finding the equation of a tangent

Example

The point A(-1, 7) lies on the curve with equation $y = 5x^2 + 2$

Find the equation of the tangent to the curve at point A.

$y = 5x^2 + 2$

$\dfrac{dy}{dx} = 10x$

so when $x = -1$

$\dfrac{dy}{dx} = -10$

Tangent has gradient -10.
Point A(-1, 7) satisfies equation of tangent.

$y - b = m(x - a)$

$y - 7 = -10(x - (-1))$

$y - 7 = -10(x + 1)$

$y - 7 = -10x - 10$

so $y = -10x - 3$

INCREASING / DECREASING FUNCTIONS

The graph below is decreasing when x is less than zero:-
The value of the function is decreasing as x is increasing.
The gradient is negative.

The graph below is increasing when x is greater than zero:-

The value of the function is increasing as x is increasing.
The gradient is positive.

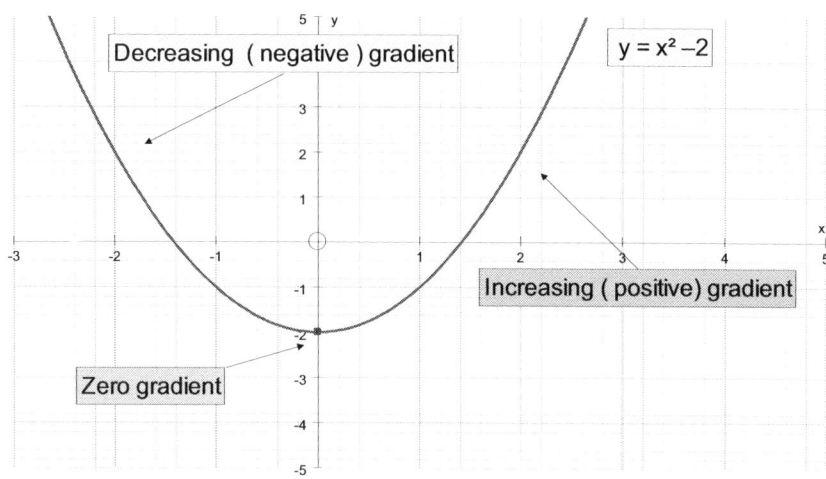

When x = 0, the gradient is zero and the graph changes from a negative gradient to a positive gradient.

This turning point is called a stationary point.
The stationary point can be a:-
- Maximum
- Minimum
- Rising point of inflection
- Falling point of inflection

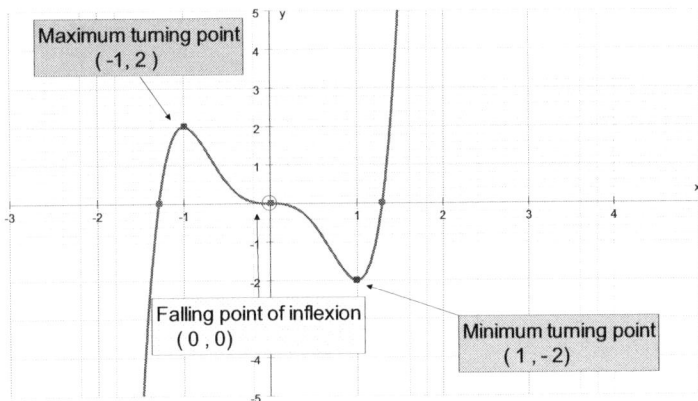

FINDING STATIONARY POINTS: REVISION

Stationary points occur when $\dfrac{dy}{dx} = 0$ (or $f'(x) = 0$)

To find the stationary points, set the first derivative of the function to zero, then factorise and solve.

Example

Find the stationary points of the graph $y = x^3 - 3x^2 + 8$

$y = x^3 - 3x^2 + 8$

$\dfrac{dy}{dx} = 3x^2 - 6x$

Stationary points occur when $\dfrac{dy}{dx} = 0$

$0 = 3x^2 - 6x$

$0 = 3x(x - 2)$

so

$3x = 0 \qquad\qquad x - 2 = 0$

$x = 0 \qquad\qquad\; x = 2$

$y = x^3 - 3x^2 + 8$

when $x = 0$ $\qquad\qquad$ when $x = 2$

$y = 0^3 - 3 \times 0^2 + 8 \qquad y = 2^3 - 3 \times 2^2 + 8$

$y = 0 - 0 + 8 \qquad\qquad\;\; y = 8 - 12 + 8$

$y = 8 \qquad\qquad\qquad\;\;\; y = 4$

Stationary points are at (0,8) and (2, 4)

NATURE TABLES: REVISION

To find out if the stationary point is a maximum, minimum or point of inflection:

Construct a nature table;
- Put in the values of x for the stationary points;
- Copy these values, with a small minus and plus sign;
- Copy the first part of the factorised form of the derivative;
- Repeat for subsequent parts of the factorised form of the derivative; and
- Write down the whole factorised form of the derivative.

Starting with the first value,
- Replace x with the value.
- Do the calculation.
- Write in the sign of the answer that you get.

Repeat for a number slightly smaller than x.
Put your answer in the small minus box.
Repeat for a number slightly larger than x.
Put your answer in the small plus box.
Multiply the signs together and put in the factorised form of derivative box.
Now draw in the tangent shape.
Repeat for other stationary points.

Nature Table of $\dfrac{dy}{dx} = 3x^2 - 6x = 3x(x-2)$

x	0 −	0	0 +	2 −	2	2 +
3x	−	0	+	+	+	+
(x−2)	−	−	−	−	0	+
$\dfrac{dy}{dx} = 3x(x-2)$	+	0	−	−	0	+
Tangent Shape	↗	→	↘	↘	→	↗
Stationary Point		Max			Min	

SECOND DERIVATIVE

The second derivative f''(x) measures the gradient with respect to x.

When this is positive, the curve is concave up – it is a minimum turning point.

When f''(x) is negative, the curve is concave down – it is a maximum turning point.

When f''(x) is zero, there may be a point of inflexion.
Draw a nature table to confirm.

$f'(x)=0$ and $f''(x)>0$	Minimum turning Point
$f'(x)=0$ and $f''(x)<0$	Maximum turning Point
$f'(x)=0$ and $f''(x)=0$	Draw table

Example

$f(x) = x^3 - 3x^2 + 8$
$f'(x) = 3x^2 - 6x$
$f''(x) = 6x - 6$

stationary points when $f'(x) = 0$
$\quad 0 = 3x^2 - 6x$
$\quad 0 = 3x(x-2)$

so $x = 0$ or $x = 2$
$f(0) = 8$ $f(2) = 8 - 12 + 8 = 4$
$f''(0) = -6$ $f''(2) = 6$
so (0,8) is a maximum   so (2,4) is a minimum

CLOSED INTERVALS

In a closed interval, the maximum and minimum values of the function occur either at a stationary point or an end point.

Example
Find the maximum and minimum values of $f(x) = x^3 - 3x^2 + 8$, $\quad -1.5 \leq x \leq 2.5$

$f(x) = x^3 - 3x^2 + 8$
$\Rightarrow f'(x) = 3x^2 - 6x$
st.pts $\quad 0 = 3x^2 - 6x$
$\Rightarrow \quad 0 = x(3x - 6)$
$\Rightarrow \quad x = 0, 2$

when $x = 0$
$f(0) = 0^3 - 3 \times 0^2 + 8$
$\Rightarrow f(0) = 8$
$f''(x) = 6x - 6$
$\Rightarrow f''(0) = -6$
Max tp at $(0, 8)$
since $f''(0) < 0$

when $x = 2$
$f(2) = 2^3 - 3 \times 2^2 + 8$
$\Rightarrow f(2) = 4$

$\Rightarrow f''(2) = 6$
Min tp at $(2, 6)$
since $f''(2) > 0$

Now check end points

$f(x) = x^3 - 3x^2 + 8$
$f(-1) = (-1.5)^3 - 3 \times (-1.5)^2 + 8$
$\quad = -3.375 - 6.75 + 8$
$\quad = -2.125$

$f(2.5) = (2.5)^3 - 3 \times (2.5)^2 + 8$
$\quad = 15.625 - 18.75 + 8$
$\quad = 4.875$

For the given interval, the maximum value is 8 and the minimum value is −2.125

HIGHER DERIVATIVES

	Newton	Liebniz
	f(x)	y = f(x)
1st Derivative	f'(x)	dy/dx
2nd Derivative	f''(x)	d²y/dx²
3rd Derivative	f'''(x)	d³y/dx³
nth Derivative	fⁿ(x)	dⁿy/dxⁿ

Example

$$y = 4x^5 - 7x^3 + 8x^2 - 3x + 4$$

$$\frac{dy}{dx} = 20x^4 - 21x^2 + 16x - 3$$

$$\frac{d^2y}{dx^2} = 80x^3 - 42x^2 + 16$$

$$\frac{d^3y}{dx^3} = 240x^2 - 84x$$

$$\frac{d^4y}{dx^4} = 480x - 84$$

$$\frac{d^5y}{dx^5} = 480$$

$$\frac{d^6y}{dx^6} = 0$$

DIFFERENTIATION CALCULATIONS REFRESHER

First Principles reminder.

$$f'(x) = \lim_{h \to 0} \frac{f(x+h) - f(x)}{h}$$

Sir Isaac Newton

$$\frac{dy}{dx} = \lim_{h \to 0} \frac{f(x+h) - f(x)}{h}$$

Baron Gottfried Von Leibniz

General rules

If $f(x) = x^n$ then $f'(x) = nx^{n-1}$ where n is a rational number

Examples

If $f(x) = ax^n$ then $f'(x) = anx^{n-1}$
a is a constant, n is a rational number

$f(x) = 2x^6$

$f'(x) = 12x^5$

$f(x) = \dfrac{3}{x} = 3x^{-1}$

$f'(x) = -3x^{-2}$

$ = -\dfrac{3}{x^2}$

$f(x) = \dfrac{4}{x} = 4x^{-1}$

$f'(x) = -4x^{-2}$

$ = -\dfrac{4}{x^2}$

$f(x) = \dfrac{1}{3x^2} = \dfrac{1}{3}x^{-2}$

$f'(x) = \dfrac{-2}{3}x^{-3}$

$ = -\dfrac{2}{3x^3}$

$f(x) = x^6$

$f'(x) = 6x^5$

$f(x) = \dfrac{1}{x} = x^{-1}$

$f'(x) = -x^{-2}$

$ = -\dfrac{1}{x^2}$

Sum Rule

$$\text{If } f(x) = g(x) + h(x)$$
$$f'(x) = g'(x) + h'(x)$$

Examples

f(x) = 4x⁶ + 3x⁷
f'(x) = 24x⁵ + 21x⁶
= 4t² (3 + 4t)

f(t) = 4t³ + 4t⁴
f'(t) = 12t² + 16t³

$$f(x) = \frac{x^2 + 5x - 1}{2\sqrt{x}}$$

$$= \frac{1}{2}\left[\frac{x^2 + 5x - 1}{\sqrt{x}}\right]$$

$$= \frac{1}{2}\left[\frac{x^2}{x^{\frac{1}{2}}} + \frac{5x}{x^{\frac{1}{2}}} - \frac{1}{x^{\frac{1}{2}}}\right]$$

$$= \frac{1}{2}\left[x^{\frac{3}{2}} + 5x^{\frac{1}{2}} - x^{-\frac{1}{2}}\right]$$

$$f(x) = 2x^2 - \frac{1}{x^2}$$
$$= 2x^2 - x^{-2}$$
$$f'(x) = 4x + 2x^{-3}$$
$$= 4x + \frac{2}{x^3}$$

$$f(x) = \frac{(x+2)^2}{x^2}$$
$$= \frac{x^2 + 4x + 4}{x^2}$$
$$= 1 + 4x^{-1} + 4x^{-2}$$
$$f'(x) = -4x^{-2} - 8x^{-3}$$
$$= \frac{-4}{x^2} - \frac{8}{x^3}$$

$$f'(x) = \frac{1}{2}\left[\frac{3}{2}x^{\frac{1}{2}} + \frac{5}{2}x^{-\frac{1}{2}} + \frac{1}{2}x^{-\frac{3}{2}}\right]$$

$$= \frac{3}{4}x^{\frac{1}{2}} + \frac{5}{4}x^{-\frac{1}{2}} + \frac{1}{4}x^{-\frac{3}{2}}$$

$$= \frac{1}{4}\left[3x^{\frac{1}{2}} + 5x^{-\frac{1}{2}} + x^{-\frac{3}{2}}\right]$$

$$= \frac{1}{4}\left[3\sqrt{x} + \frac{5}{\sqrt{x}} + \frac{x}{\sqrt{x^3}}\right]$$

$$f(x) = (x+1)(x+3)$$
$$= x^2 + 4x + 3$$
$$f'(x) = 2x + 4$$

$$\boxed{\text{If } f(x) = (x + a)^n \text{ then } f'(x) = n(x + a)^{n-1}}$$
a is a real number, n is a rational number

Examples

$f(x) = (x+5)^7$

$f'(x) = 7(x+5)^6$

$f(x) = \dfrac{1}{x+3}$

$= (x+3)^{-1}$

$f'(x) = -1(x+3)^{-2}$

$f'(x) = \dfrac{-1}{(x+3)^2}$

$f(x) = \sqrt[5]{(x+3)^2}$

$= (x+3)^{2/5}$

$f'(x) = \dfrac{2}{5}(x+3)^{-3/5}$

$$\boxed{\text{If } f(x) = (ax + b)^n \text{ then } f'(x) = an(ax + b)^{n-1}}$$
a is a real number, n is a rational number

Examples

$f(x) = (5x+5)^7$

$f'(x) = 35(5x+5)^6$

$f(x) = \dfrac{1}{3x+3}$

$= (3x+3)^{-1}$

$f'(x) = -3(3x+3)^{-2}$

$f'(x) = \dfrac{-3}{(3x+3)^2}$

$f(x) = \sqrt[3]{(4x+3)^2}$

$= (4x+3)^{2/3}$

$f'(x) = \dfrac{8}{3}(4x+3)^{-1/3}$

THE CHAIN RULE

$$h'(x) = g'(f(x)) \times f'(x)$$ or in Leibnitz notation $$\frac{dy}{dx} = \frac{dy}{du} \times \frac{du}{dx}$$

Example

$h(x) = (x^9 + 8x)^3$
let $h(x) = g(f(x))$
where $f(x) = x^9 + 8x$ and $g(x) = x^3$
$f'(x) = 9x^8 + 8$
$g'(x) = 3x^2$

In Leibnitz notation
$y = (x^9 + 8x)^3$
Let $u = x^9 + 8x$ then $y = u^3$
$\frac{du}{dx} = 9x^8 + 8$ and $\frac{dy}{du} = 3u^2$

$h'(x) = g'(f(x)) \times f'(x)$
$h'(x) = 3(x^9 + 8x)^2 \times (9x^8 + 8)$
$h'(x) = 3(9x^8 + 8)(x^9 + 8x)^2$

$\frac{dy}{dx} = \frac{dy}{du} \times \frac{du}{dx}$
$= 3u^2 \times (9x^8 + 8)$
$= 3(x^9 + 8x)^2 \times (9x^8 + 8)$
$= 3(9x^8 + 8)(x^9 + 8x)^2$

Longer questions use a continuation

Example

$$\frac{dy}{dx} = \frac{dy}{du} \times \frac{du}{dt} \times \frac{dt}{dx}$$

Differentiate $y = \sin^3(4x + 5)$
Let $y = u^3$, $u = \sin t$ and $t = 4x + 5$
then $\frac{dy}{du} = 3u^2$, $\frac{du}{dt} = \cos t$, $\frac{dt}{dx} = 4$

$\frac{dy}{dx} = \frac{dy}{du} \times \frac{du}{dt} \times \frac{dt}{dx}$
$= 3u^2 \times \cos t \times 4$
$= 12(\sin t)^2 \times \cos(4x + 5)$
$= 12\sin^2(4x + 5)\cos(4x + 5)$

PRODUCT RULE

$$\text{If } f(x) = g(x).h(x) \text{ then } f'(x) = g'(x).h(x) + g(x).h'(x)$$

Examples

$f(x) = x^3 \cos x$

$f'(x) = 3x^2 \cdot \cos x + x^3 \cdot (-\sin x)$

$\quad = 3x^2 \cos x - x^3 \sin x$

$\quad = x^2 (3\cos x - x\sin x)$

$f(x) = (x+1)^3 (x-1)^2$

$f'(x) = 3(x+1)^2 \cdot 1 \cdot (x-1)^2 + (x+1)^3 \cdot 2(x-1) \cdot 1$

$\quad = 3(x+1)^2 (x-1)^2 + (x+1)^3 \times 2(x-1)$

$\quad = 3(x+1)^2 (x-1)^2 + 2(x+1)^3 (x-1)$

$\quad = (x+1)(x-1)\left(3(x+1)(x-1) + 2(x+1)^2\right)$

$f(x) = x^2 (x+1)^3 (x-1)^2$

$f'(x) = 2x(x+1)^3 (x-1)^2 + 3(x+1)^2 x^2 (x-1)^2 + 2(x-1)x^2 (x+1)^3$

$\quad = (x+1)^2 (x-1)x\left(2(x+1)(x-1) + 3x(x-1) + 2x(x+1)\right)$

Don't forget the chain rule!

$f(x) = x^2 (5x+1)^3 (2x-1)^2$

$f'(x) = 2x(5x+1)^3 (2x-1)^2 + 3(5x+1)^2 \bullet 5 \bullet x^2 (2x-1)^2 + 2(2x-1) \bullet 2 \bullet x^2 (5x+1)^3$

$\quad = 2x(5x+1)^3 (2x-1)^2 + 15(5x+1)^2 x^2 (2x-1)^2 + 4(2x-1)x^2 (5x+1)^3$

$\quad = (5x+1)^2 (2x-1)x\left(2(5x+1)(2x-1) + 15x(2x-1) + 4x(5x+1)\right)$

QUOTIENT RULE

$$\text{If} \quad f(x) = \frac{g(x)}{h(x)}$$

$$\text{then,} \quad f'(x) = \left(\frac{g(x)}{h(x)}\right)' = \frac{g'(x)h(x) - g(x)h'(x)}{(h(x))^2}$$

Example

$$f(x) = \frac{x^4}{x+1}, \quad \text{find } f'(x)$$

$$f'(x) = \frac{4x^3 \cdot (x+1) - x^4 \cdot 1}{(x+1)^2}$$

$$= \frac{4x^3(x+1) - x^4}{(x+1)^2}$$

$$= \frac{4x^4 + 4x^3 - x^4}{(x+1)^2}$$

$$= \frac{x^3(3x+4)}{(x+1)^2}$$

TRIG FUNCTIONS

$$\boxed{\text{If } f(x) = \sin ax \text{ then } f'(x) = a\cos ax}$$

Example

f(x) = 8sinx   f(x) = sin2x   f(x) = 3sin5x
f'(x) = 8cosx   f'(x) = 2cos2x   f'(x) = 15cos5x

$$\boxed{\text{If } f(x) = \cos ax \text{ then } f'(x) = -a\sin ax}$$

Example

f(x) = 8cosx   f(x) = 2cos5x   f(x) = 3cos1/3x
f'(x) = -8sinx   f'(x) = -10sin5x f'(x) = -sin1/3x

$$\boxed{\text{If } f(x) = \tan x \text{ then } f'(x) = \frac{1}{\cos^2 x} = \sec^2 x}$$

Example

$y = \tan^2 4x$, find dy/dx

$y = \tan^2 4x$

let $y = u^2$ $u = \tan 4x$

$dy/du = 2u$ $du/dx = \sec^2(4x) \cdot 4$

$dy/dx = 2u \cdot 4\sec^2(4x)$
$ = 8\tan(4x)\sec^2(4x)$

$$\boxed{\text{If } f(x) = \operatorname{cosec} x \text{ then } f'(x) = -\operatorname{cosec} x \cot x}$$

Example

$y = \operatorname{cosec}(2x+3)$, find dy/dx

$$\frac{dy}{dx} \operatorname{cosec}(2x+3) = -\operatorname{cosec}(2x+3)\cot(2x+3) \cdot 2$$
$$= -2\operatorname{cosec}(2x+3)\cot(2x+3)$$

$$\boxed{\text{If } f(x) = \sec x \text{ then } f'(x) = \sec x \tan x}$$

Example

$y = \sec(4 - 3x^2)$, find dy/dx

$$\frac{dy}{dx} \sec(4-3x^2) = \sec(4-3x^2)\tan(4-3x^2) \cdot -6x$$
$$= -6x\sec(4-3x^2)\tan(4-3x^2)$$

$$\boxed{\text{If } f(x) = \cot x \text{ then } f'(x) = -\operatorname{cosec}^2 x}$$

Example

$y = \cot(5x^3)$, find dy/dx

$$\frac{dy}{dx} \cot(5x^3) = -\operatorname{cosec}^2(5x^3) \cdot 15x^2$$
$$= -15x^2 \operatorname{cosec}^2(5x^3)$$

LOGARITHMS AND EXPONENTIALS

$$\boxed{\text{If } f(x) = e^x \text{ then } f'(x) = e^x}$$

Examples

$y = e^{9x^2}$ find dy/dx

$\dfrac{dy}{dx} = e^{9x^2} \cdot 18x$

$= 18x e^{9x^2}$

$y = e^x \cos x$ find dy/dx

$\dfrac{dy}{dx} = e^x \cdot \cos x + e^x \cdot -\sin x$

$= e^x \cos x - e^x \sin x$

$= e^x (\cos x - \sin x)$

$$\boxed{\text{If } f(x) = \log_e x \text{ then } f'(x) = \dfrac{1}{x}}$$

Examples

$y = \ln(5x^2 - 6)$ find dy/dx

$\dfrac{dy}{dx} \ln(5x^2 - 6) = \dfrac{1}{5x^2 - 6} \cdot 10x$

$= \dfrac{10x}{5x^2 - 6}$

$y = (\ln x)^2 e^x$ find dy/dx

$\dfrac{dy}{dx} = 2\ln x \cdot \dfrac{1}{x} \cdot e^x + (\ln x)^2 e^x$

$= \dfrac{2\ln x}{x} e^x + (\ln x)^2 e^x$

$= e^x \ln x \left(\dfrac{2}{x} + \ln x \right)$

$y = \dfrac{\ln(\cos x)}{x^2}$ find dy/dx

Use Quotient rule

Don't forget the Chain rule !

$\dfrac{dy}{dx} = \dfrac{\dfrac{1}{\cos x} \cdot -\sin x \cdot x^2 - 2x \ln(\cos x)}{x^4}$

$= \dfrac{\dfrac{-\sin x}{\cos x} x^2 - 2x \ln(\cos x)}{x^4}$

$= \dfrac{-x^2 \tan x - 2x \ln(\cos x)}{x^4}$

$= \dfrac{-x \tan x - 2\ln(\cos x)}{x^3}$

212

APPROXIMATING ROOTS OF AN EQUATION : NEWTON'S METHOD

$$x_{n+1} = x_n - \frac{f(x_n)}{f'(x_n)}$$

Find the roots of the function

$y = 12x^3 + 4x^2 - 15x - 4$

$f'(x) = 36x^2 + 8x - 15$

$$x_{n+1} = x_n - \frac{f(x_n)}{f'(x_n)}$$

$$x_{n+1} = x_n - \frac{12x^3 + 4x^2 - 15x - 4}{36x^2 + 8x - 15}$$

try

$x_n = 1.5$

$$x_{n+1} = 1.5 - \frac{12(1.5)^3 + 4(1.5)x^2 - 15(1.5) - 4}{36(1.5)^2 + 8(1.5)x - 15} = 1.205128205$$

This is then put back into the right hand side equation.
Continue until the answer is found.

X_{n+1}	Xn	f(x)	f'(x)	f(x)/f'(x)
1.205128	1.5	23	78	0.294872
1.104214	1.205128	4.735397	46.92505	0.100914
1.091751	1.104214	0.470216	37.72811	0.012463
1.091566	1.091751	0.006773	36.64313	0.000185
1.091566	1.091566	1.48E-06	36.62712	4.04E-08
1.091566	1.091566	7.11E-14	36.62712	1.94E-15

APPLICATIONS OF DIFFERENTIATION

Finding greatest/least values.

Example

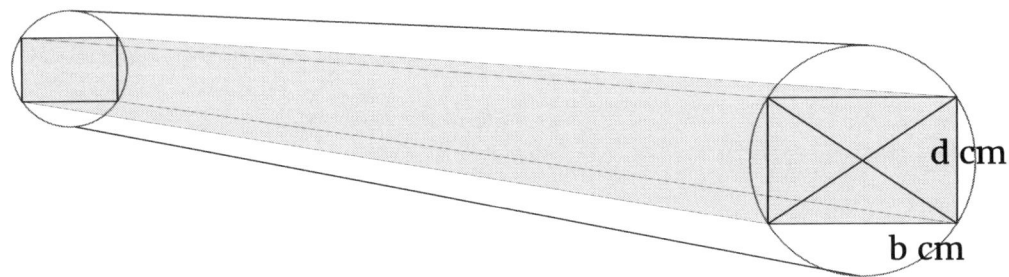

A rectangular beam is to be cut out of a cylinder.

The diameter of the cylinder is 40cm.
The breadth of the beam is b cm.
The depth of the beam is d cm.

The strength of the beam is given by the formula $S = 1.7b(400 - b^2)$

What dimensions of the beam are required for the beam to have maximum strength?

$S = 1.7b(400 - b^2)$
$\quad = 680b - 1.7b^3$
$\dfrac{dS}{db} = 680 - 5.1b^2$

The maximum will occur when $\dfrac{dS}{db} = 0$

$0 = 680 - 5.1b^2$
$\Rightarrow 5.1b^2 = 680$
$\Rightarrow b^2 = \dfrac{680}{5.1} = 133\dfrac{1}{3} = \dfrac{400}{3}$
$\Rightarrow b = \sqrt{\dfrac{400}{3}}$
$\Rightarrow b = \dfrac{20}{\sqrt{3}}$

Check that this is a maximum

$f'(S) = 680 - 5.1b^2$

f'(S)	$19 \div \sqrt{3}$	$20 \div \sqrt{3}$	$21 \div \sqrt{3}$
	+	0	−

Maximum occurs when b = 20 ÷ √3 cm

Find d

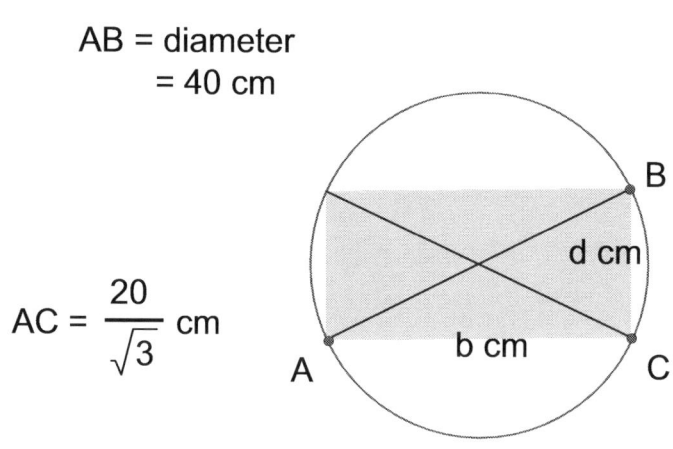

AB = diameter = 40 cm

AC = $\dfrac{20}{\sqrt{3}}$ cm

So d ≅ 38.30 cm (2dp)
and b ≅ 11.55 cm (2dp)

By pythagoras' theorem
$AB^2 = AC^2 + BC^2$

$\Rightarrow 40^2 = \left(\dfrac{20}{\sqrt{3}}\right)^2 + BC^2$

$\Rightarrow BC^2 = 40^2 - \left(\dfrac{20}{\sqrt{3}}\right)^2$

$\Rightarrow BC^2 = 40^2 - \dfrac{400}{3}$

$\Rightarrow BC^2 = 40(40 - \dfrac{10}{3})$

$\Rightarrow BC^2 = 40 \times \dfrac{110}{3}$

$\Rightarrow BC^2 = \dfrac{400 \times 11}{3}$

$\Rightarrow BC = \sqrt{\dfrac{400 \times 11}{3}}$

$\Rightarrow BC = 20\sqrt{\dfrac{11}{3}}$ cm

RECTILINEAR MOTION (STRAIGHT LINE MOTION)

If an object moves in a straight line along the x-axis, then after t seconds it has moved a distance s unit from the origin.   s = f(t)

velocity = rate of change of displacement with time

$$v = \frac{ds}{dt}$$

acceleration = rate of change of velocity with time

$$a = \frac{dv}{dt} = \frac{d^2s}{dt^2}$$

Newton's equations of Motion:-

$$v = u + at$$
$$s = ut + \frac{1}{2}at^2$$
$$v^2 = u^2 + 2as$$

$v =$ *final velocity* $u =$ *initial velocity*
$s =$ *distance travelled*
$a =$ *acceleration* $t =$ *time taken*

Example

A ball is thrown vertically upwards.

The height reached by the ball after t seconds is $h = 6t - t^2$.

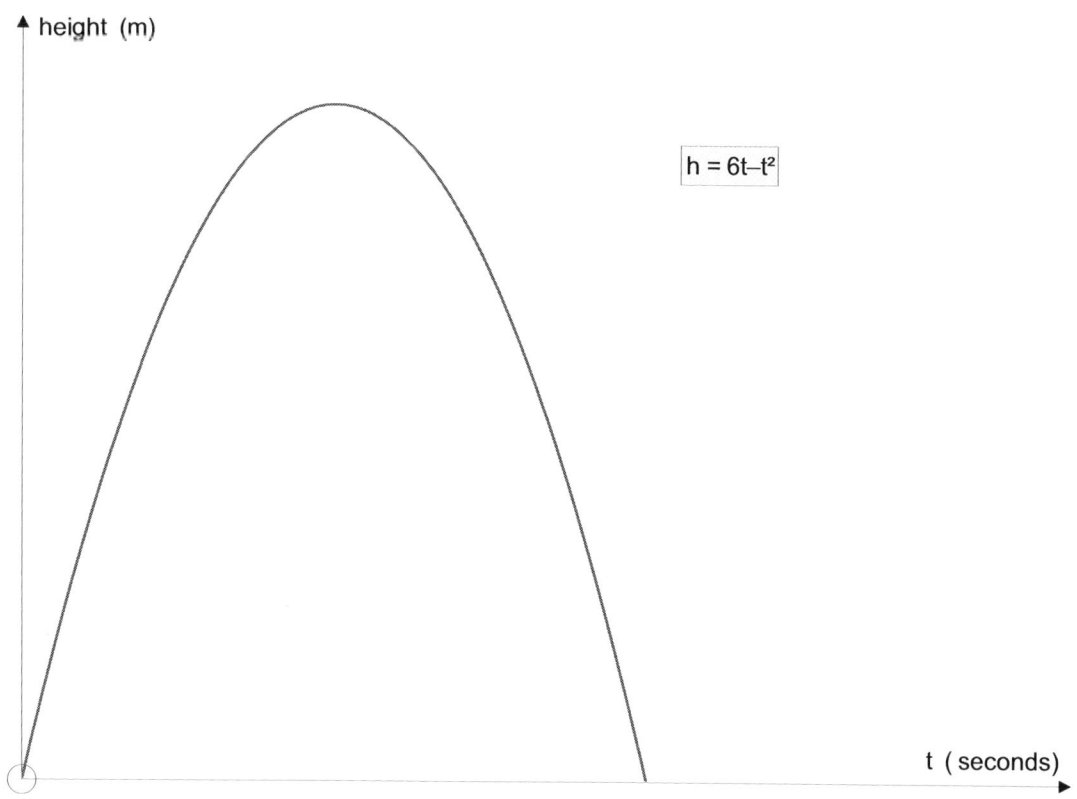

a) How long does the ball take to reach its maximum height?

b) What is the maximum height?

c) What is the velocity of the ball 5 seconds after being thrown?

d) What is the velocity of the ball when it hits the ground?

e) Is the ball accelerating or decelerating at this point?

$h = 6t - t^2$

$\dfrac{dh}{dt} = 6 - 2t$

a) Maximum height occurs when $\dfrac{dh}{dt} = 0$,

since velocity at this point is zero

$0 = 6 - 2t$

$\Rightarrow 2t = 6$

$\Rightarrow t = 3$

The maximum height reached occurs after 3 seconds.

b) $h = 6t - t^2$
$= 6 \times 3 - 3^2$
$= 18 - 3^2$
$= 9m$

The maximum height reached is 9m.

c) After 5 seconds

$v = f'(h) = 6 - 2t$

$v = f'(5) = 6 - 10$

$= -4 \, m/s$

The velocity of the ball is -4 m/s

d) The ball hits the ground when h = 0

This occurs when $0 = 6t - t^2$

i.e $0 = t(6 - t)$

$t = 0$ or $t = 6$ seconds

$v = f'(h) = 6 - 2t$

$v = f'(6) = 6 - 12$

$= -6 \, m/s$

The velocity of the ball is -6 m/s

e) The ball hits the ground when h = 0

$v = f'(h) = 6 - 2t$

$a = f''(h) = -2$

The ball maintains a constant deceleration (-2 m/s^2) throughout its trip.

DIFFERENTIATING INVERSE FUNCTIONS

Find the inverse of the function.

Differentiate the function.

Plug data into following equation:-

$$\frac{d}{dx}f^{-1}(x) = \frac{1}{f'(f^{-1}(x))}$$

Examples

Given $f(x) = x^6$, find f'(x) and state the derivative of $f^{-1}(x)$.

$f(x) = x^6$ $\qquad\qquad\qquad\qquad$ $y = x^6$

$f'(x) = 6x^5$ $\qquad\qquad\qquad\qquad$ $\sqrt[6]{y} = x$

$f^{-1}(x) = \sqrt[6]{x}$ $\qquad\qquad\qquad$ Inverse $y = \sqrt[6]{x}$

$$\frac{d}{dx}f^{-1}(x) = \frac{1}{f'(f^{-1}(x))}$$

$$= \frac{1}{6(\sqrt[6]{x})^5}$$

$$= \frac{1}{6x^{5/6}}$$

Given $f(x) = 3x^{-2}$, find $f'(x)$ and state the derivative of $f^{-1}(x)$.

$f(x) = 3x^{-2}$

$f'(x) = -6x^{-3}$

$f^{-1}(x) = \left(\dfrac{3}{x}\right)^{1/2}$

$y = 3x^{-2}$

$y = \dfrac{3}{x^2}$

$x^2 = \dfrac{3}{y}$

$x = \sqrt{\dfrac{3}{y}}$

Inverse $y = \sqrt{\dfrac{3}{x}}$

$\dfrac{d}{dx} f^{-1}(x) = \dfrac{1}{f'(f^{-1}(x))}$

$= \dfrac{1}{-6\left(\left(\dfrac{3}{x}\right)^{1/2}\right)^{-3}}$

$= \dfrac{1}{-6\left(\dfrac{3}{x}\right)^{-3/2}}$

$= \dfrac{1}{-6 \times \dfrac{1}{\left(\dfrac{3}{x}\right)^{3/2}}}$

$= \dfrac{1}{-6} \times \left(\dfrac{3}{x}\right)^{3/2}$

$= \dfrac{\sqrt{27}}{-6x^{3/2}}$

$= \dfrac{3\sqrt{3}}{-6x^{3/2}} = \dfrac{\sqrt{3}}{-2x^{3/2}}$

Note

$\dfrac{\sqrt{3}}{-2x^{3/2}} \times \dfrac{\sqrt{3}}{\sqrt{3}} = \dfrac{3}{-2\sqrt{3}x^{3/2}}$

$= \dfrac{3}{-2\sqrt{3} \times x\sqrt{x}}$

$= \dfrac{3}{-2x \times \sqrt{3} \times \sqrt{x}}$

$= \dfrac{3}{-2x\sqrt{3x}}$

Express $f(x) = x^2 + 2x + 4$, $x \geq 4$, in the form $p(x+q)^2 + r$.
Find $f'(x)$ and state the derivative of $f^{-1}(x)$.

$f(x) = x^2 + 2x + 4$
$= (x+1)^2 + 4 - 1$
$= (x+1)^2 + 3$

$f'(x) = 2(x+1)$

$f^{-1}(x) = \sqrt{x-3} - 1$

$y = (x+1)^2 + 3$
$y - 3 = (x+1)^2$
$\sqrt{y-3} = x+1$
$x = \sqrt{y-3} - 1$
Inverse $y = \sqrt{x-3} - 1$

$\dfrac{d}{dx} f^{-1}(x) = \dfrac{1}{f'(f^{-1}(x))}$

$= \dfrac{1}{2\left(\left(\sqrt{x-3} - 1\right) + 1\right)}$

$= \dfrac{1}{2\left(\sqrt{x-3}\right)}$

$= \dfrac{1}{2\sqrt{x-3}}$

$y = f(x)$
$\Rightarrow x = f^{-1}(y)$

$\dfrac{dx}{dy} = \dfrac{1}{f'(f^{-1}(y))}$
$= \dfrac{1}{f'(x)}$
$= \dfrac{1}{\dfrac{dy}{dx}}$

$\dfrac{dx}{dy} = \dfrac{1}{\dfrac{dy}{dx}}$

Express $f(x) = 4x^2 + 2x + 4$, $x \geq 4$, in the form $p(x + q)^2 + r$.

Find f'(x) and state the derivative of $f^{-1}(x)$.

$f(x) = 4x^2 + 2x + 4$
$= 4(x^2 + 1/2x + 1)$
$= 4((x + 1/4)^2 + 1 - 1/16)$
$= 4((x + 1/4)^2 + 15/16)$
$= 4(x + 1/4)^2 + 15/4$

$y = 4(x + 1/4)^2 + 15/4$

$\sqrt{\dfrac{y - 15/4}{4}} = x + 1/4$

$f'(x) = 8(x + 1/4)$

$x = \sqrt{\dfrac{y - 15/4}{4}} - 1/4$

$f^{-1}(x) = \dfrac{1}{2}\sqrt{\left(x - \dfrac{15}{4}\right)} - 1/4$

$x = \sqrt{\dfrac{1}{4}\left(y - \dfrac{15}{4}\right)} - 1/4$

$x = \dfrac{1}{2}\sqrt{\left(y - \dfrac{15}{4}\right)} - 1/4$

Inverse $y = \dfrac{1}{2}\sqrt{\left(x - \dfrac{15}{4}\right)} - 1/4$

$\dfrac{d}{dx} f^{-1}(x) = \dfrac{1}{f'(f^{-1}(x))}$

$= \dfrac{1}{8\left(\left(\dfrac{1}{2}\sqrt{\left(x - \dfrac{15}{4}\right)} - 1/4\right) + 1/4\right)}$

$= \dfrac{1}{4\sqrt{\left(x - \dfrac{15}{4}\right)}}$

DIFFERENTIATING INVERSE TRIG FUNCTIONS

Given $f(x) = \sin x$ for the interval $-\dfrac{\pi}{2} \leq x \leq \dfrac{\pi}{2}$, find the derivative of $f^{-1}(x)$.

$f(x) = \sin x$

$f'(x) = \cos x$

$f^{-1}(x) = \sin^{-1}(x)$ 　　Let $\sin^{-1}(x) = \theta$

　　　　　　　　　　　　　Then $x = \sin \theta$

The given interval allows the existance of a right angled triangle with opposite side length x and hypotenuse 1.

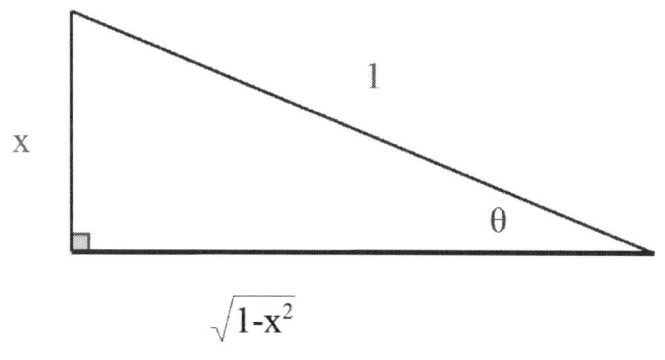

$\dfrac{d}{dx} f^{-1}(x) = \dfrac{1}{f'(f^{-1}(x))}$

$= \dfrac{1}{\cos(\sin^{-1}(x))}$

$= \dfrac{1}{\cos(\theta)}$

$= \dfrac{1}{\left(\dfrac{\sqrt{1-x^2}}{1}\right)}$

$= \dfrac{1}{\sqrt{1-x^2}}$

$$\boxed{\dfrac{d}{dx} \sin^{-1}(x) = \dfrac{1}{\sqrt{1-x^2}}}$$

Given f(x) = cosx for the interval $-\frac{\pi}{2} \leq x \leq \frac{\pi}{2}$, find the derivative of $f^{-1}(x)$.

$f(x) = \cos x$
$f'(x) = -\sin x$
$f^{-1}(x) = \cos^{-1}(x)$ Let $\cos^{-1}(x) = \theta$
Then $x = \cos \theta$

The given interval allows the existance of a right angled triangle with adjacent side length x and hypotenuse 1.

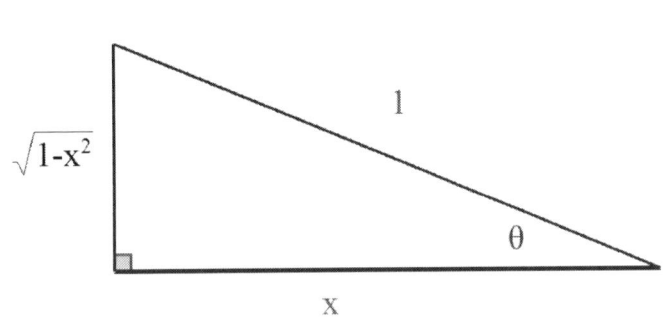

$\frac{d}{dx} f^{-1}(x) = \frac{1}{f'(f^{-1}(x))}$

$= \frac{1}{-\sin(\cos^{-1}(x))}$

$= \frac{1}{-\sin(\theta)}$

$= \frac{-1}{\left(\frac{\sqrt{1-x^2}}{1}\right)}$

$= \frac{-1}{\sqrt{1-x^2}}$

$$\boxed{\frac{d}{dx} \cos^{-1}(x) = \frac{-1}{\sqrt{1-x^2}}}$$

Given f(x) = tanx for the interval $-\dfrac{\pi}{2} < x < \dfrac{\pi}{2}$, find the derivative of $f^{-1}(x)$.

$f(x) = \tan x$

$f'(x) = \sec^2 x$

$f^{-1}(x) = \tan^{-1}(x)$ Let $\tan^{-1}(x) = \theta$

Then $x = \tan\theta$

The given interval allows the existance of a right angled triangle with opposite side length x and adjacent length 1.

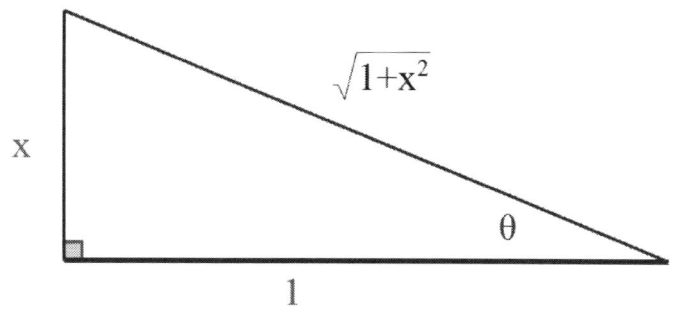

$$\dfrac{d}{dx}f^{-1}(x) = \dfrac{1}{f'(f^{-1}(x))}$$

$$= \dfrac{1}{\sec^2(\tan^{-1}(x))}$$

$$= \dfrac{1}{\sec^2(\theta)}$$

$$= \cos^2(\theta)$$

$$= \left(\dfrac{1}{\sqrt{x^2+1}}\right)^2$$

$$= \dfrac{1}{x^2+1}$$

$$\boxed{\dfrac{d}{dx}\tan^{-1}(x) = \dfrac{1}{1+x^2}}$$

Example

Find $\dfrac{d}{dx}(\tan^{-1} x^2)$

$y = \tan^{-1} x^2$

$$\begin{aligned}\dfrac{dy}{dx} &= \dfrac{d}{dx}\tan^{-1} x^2 \bullet 2x \\ &= \dfrac{1}{1+\left(x^2\right)^2} \bullet 2x \\ &= \dfrac{2x}{1+x^4}\end{aligned}$$

Don't forget the chain rule!!

DIFFERENTIATING EXPLICIT AND IMPLICIT FUNCTIONS

An explicit function is one which is given in terms of the independent variable.

Take the following function,

$y = x^2 + 3x - 8$

y is the dependent variable and is given in terms of the independent variable x.

Note that y is the subject of the formula.

Implicit functions, on the other hand, are usually given in terms of both dependent and independent variables.

eg:- $y + x^2 - 3x + 8 = 0$

Sometimes, it is not convenient to express a function explicitly.
For example, the circle $x^2 + y^2 = 16$ could be written as

$y = \sqrt{16 - x^2}$ or $y = -\sqrt{16 - x^2}$

Which version should be taken if the function is to be differentiated?

It is often easier to differentiate an implicit function without having to rearrange it, by differentiating each term in turn.

> Since y is a function of x, the chain, product and quotient rules apply!

Examples

Differentiate $x^2 + y^2 = 16$ with respect to x.

$$2x + 2y \frac{dy}{dx} = 0$$

$$\Rightarrow 2y \frac{dy}{dx} = -2x$$

$$\Rightarrow \frac{dy}{dx} = \frac{-2x}{2y}$$

$$\Rightarrow \frac{dy}{dx} = \frac{-x}{y}$$

Compared to

$$x^2 + y^2 = 16$$
$$\Rightarrow y^2 = 16 - x^2$$
$$\Rightarrow y = \sqrt{16 - x^2}$$
$$\Rightarrow y = (16 - x^2)^{1/2}$$
$$\Rightarrow \frac{dy}{dx} = \frac{1}{2}(16 - x^2)^{-1/2} \cdot -2x$$
$$\Rightarrow \frac{dy}{dx} = -x(16 - x^2)^{-1/2}$$
$$\Rightarrow \frac{dy}{dx} = \frac{-x}{\sqrt{16 - x^2}}$$
$$\Rightarrow \frac{dy}{dx} = \frac{-x}{y}$$

Differentiate $2x^2 + 2xy + 2y^2 = 16$ with respect to x.

$$2x^2 + 2xy + 2y^2 = 16$$

$$\Rightarrow 4x + 2\left(1 \cdot y + x \frac{dy}{dx}\right) + 2 \cdot 2y \cdot \frac{dy}{dx} = 0$$

$$\Rightarrow 4x + 2\left(y + x \frac{dy}{dx}\right) + 4y \frac{dy}{dx} = 0$$

$$\Rightarrow 4x + 2y + \frac{dy}{dx}(2x + 4y) = 0$$

$$\Rightarrow \frac{dy}{dx}(2x + 4y) = -4x - 2y$$

$$\Rightarrow \frac{dy}{dx} = \frac{-2(2x + y)}{(2x + 4y)}$$

$$\Rightarrow \frac{dy}{dx} = \frac{-(2x + y)}{(x + 2y)}$$

Find the gradient of the tangent at the point R(1,2) on the graph of the curve defined by $x^3+y^2=5$, and determine whether the curve is concave up or concave down at this point.

$x^3+y^2=5$

$\Rightarrow 3x^2+2y\dfrac{dy}{dx}=0$

$\Rightarrow \dfrac{dy}{dx}=\dfrac{-3x^2}{2y}$

At R(1,2)

$\Rightarrow \dfrac{dy}{dx}=\dfrac{-3}{4}$

$3x^2+2y\dfrac{dy}{dx}=0$

$\Rightarrow 6x+2\dfrac{dy}{dx}\cdot\dfrac{dy}{dx}+2y\dfrac{d^2y}{dx^2}=0$

$\Rightarrow 6x+2\left(\dfrac{-3x^2}{2y}\right)\cdot\left(\dfrac{-3x^2}{2y}\right)+2y\dfrac{d^2y}{dx^2}=0$

$\Rightarrow 6x+2\left(\dfrac{9x^4}{4y^2}\right)+2y\dfrac{d^2y}{dx^2}=0$

$\Rightarrow 6x+\dfrac{9x^4}{2y^2}+2y\dfrac{d^2y}{dx^2}=0$

$\Rightarrow 2y\dfrac{d^2y}{dx^2}=-6x-\dfrac{9x^4}{2y^2}$

$\Rightarrow \dfrac{d^2y}{dx^2}=\dfrac{-3x}{y}-\dfrac{9x^4}{4y^3}$

$\Rightarrow \dfrac{d^2y}{dx^2}=\dfrac{-12xy^3-9x^4y}{4y^4}$

$\Rightarrow \dfrac{d^2y}{dx^2}=\dfrac{-3xy(4y^2+3x^3)}{4y^4}$

$\Rightarrow \dfrac{d^2y}{dx^2}=\dfrac{-3x(4y^2+3x^3)}{4y^3}$

At R(1,2)

$\dfrac{d^2y}{dx^2}=\dfrac{-3(4\times 2^2+3)}{4\times 2^3}=\dfrac{-3\times 19}{32}$

∴ the curve is concave down

since $\dfrac{d^2y}{dx^2}<0$

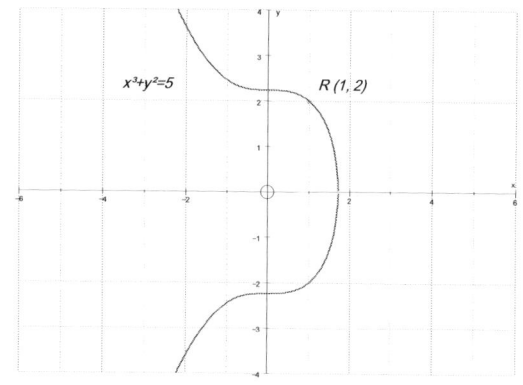

LOGARITHMIC DIFFERENTIATION

Using logs to differentiate nastier functions.

Examples

Differentiate 5^{2x}

$f(x) = 5^{2x}$
let $y = 5^{2x}$ Take logs of both sides
$\ln y = \ln 5^{2x}$
$\Rightarrow \ln y = 2x \ln 5$
$\Rightarrow \dfrac{1}{y}\dfrac{dy}{dx} = 2\ln 5$
$\Rightarrow \dfrac{dy}{dx} = y \, 2\ln 5$
$\Rightarrow \dfrac{dy}{dx} = 5^{2x} 2\ln 5$ (Substituting back for y)
$\Rightarrow \dfrac{dy}{dx} = 5^{2x} \ln 25$

Example

Differentiate $y = \dfrac{(x+1)^{\frac{1}{5}}(x+1)^{\frac{1}{3}}}{(x+2)^{\frac{1}{4}}}$

$$y = \frac{(x+1)^{\frac{1}{5}}(x+1)^{\frac{1}{3}}}{(x+2)^{\frac{1}{4}}}$$

$$\ln y = \ln(x+1)^{\frac{1}{5}} + \ln(x+1)^{\frac{1}{3}} - \ln(x+2)^{\frac{1}{4}}$$

$$\Rightarrow \ln y = \frac{1}{5}\ln(x+1) + \frac{1}{3}\ln(x+1) - \frac{1}{4}\ln(x+2)$$

$$\Rightarrow \frac{1}{y}\frac{dy}{dx} = \frac{1}{5(x+1)} + \frac{1}{3(x+1)} - \frac{1}{4(x+2)}$$

$$\Rightarrow \frac{1}{y}\frac{dy}{dx} = \frac{8}{15(x+1)} - \frac{1}{4(x+2)}$$

$$\Rightarrow \frac{1}{y}\frac{dy}{dx} = \frac{32(x+2) - 15(x+1)}{60(x+1)(x+2)}$$

$$\Rightarrow \frac{1}{y}\frac{dy}{dx} = \frac{17x+49}{60(x+1)(x+2)}$$

$$\Rightarrow \frac{dy}{dx} = y\frac{17x+49}{60(x+1)(x+2)}$$

$$\Rightarrow \frac{dy}{dx} = \frac{(x+1)^{\frac{1}{5}}(x+1)^{\frac{1}{3}}}{(x+2)^{\frac{1}{4}}} \times \frac{(17x+49)}{60(x+1)(x+2)}$$

$$\Rightarrow \frac{dy}{dx} = \frac{(x+1)^{\frac{1}{5}}(x+1)^{\frac{1}{3}}(17x+49)}{60(x+1)(x+2)^{\frac{5}{4}}}$$

$$\Rightarrow \frac{dy}{dx} = \frac{(x+1)^{\frac{8}{15}}(17x+49)}{60(x+1)(x+2)^{\frac{5}{4}}}$$

$$\Rightarrow \frac{dy}{dx} = \frac{(17x+49)}{60(x+1)^{\frac{7}{15}}(x+2)^{\frac{5}{4}}}$$

PARAMETRIC EQUATIONS

If f(t) and g(t) are functions of t, then x=f(t) and y=g(t), define a point on a curve for each value of t.

x=f(t) and y=g(t) are called parametric equations, whilst t is the parameter.

Examples

The curve defined by
X = 3(1+t)
y = 6(1-t)
looks like this ->

at t =0, x = 3 and y= 6, giving the point (3,6).

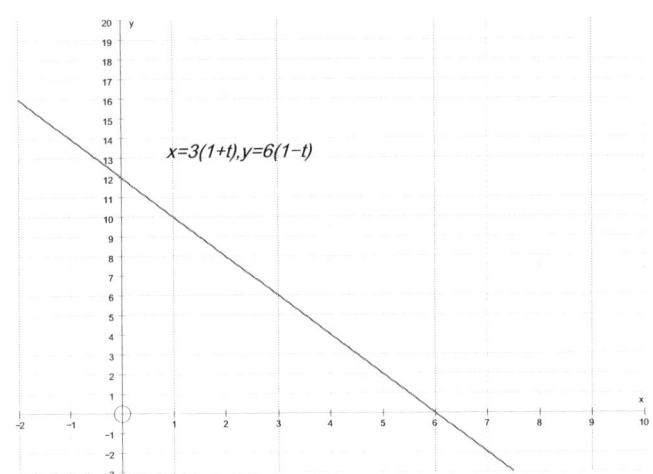

A circle can be described by the parametric equations

x = rcost
y = rsint   where r is the radius of the circle with centre (0,0).

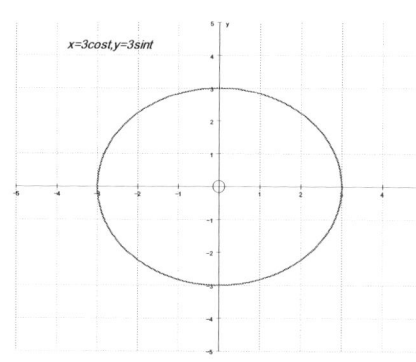

Example

x = 3cost
y = 3sint

represents a circle of radius 3, centre (0,0).

Where as

x = 3 + 3cost
y = 5 + 3sint

represents a circle of radius 3, centre

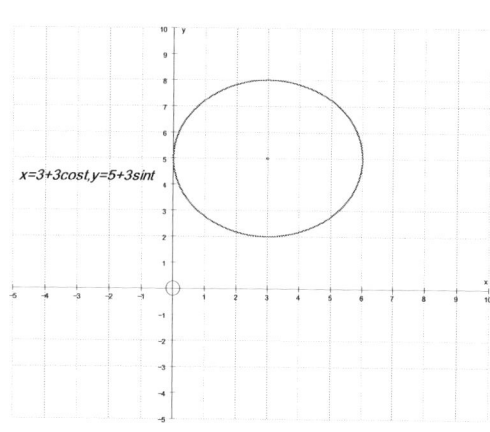

CONSTRAINT EQUATIONS

(Or how to try to get back to y=f(x))

Given x=f(t), t=f⁻¹(x), so y=g(t) become

Example

Find the constraint equation of the para

x=4-t
y=1+3t.

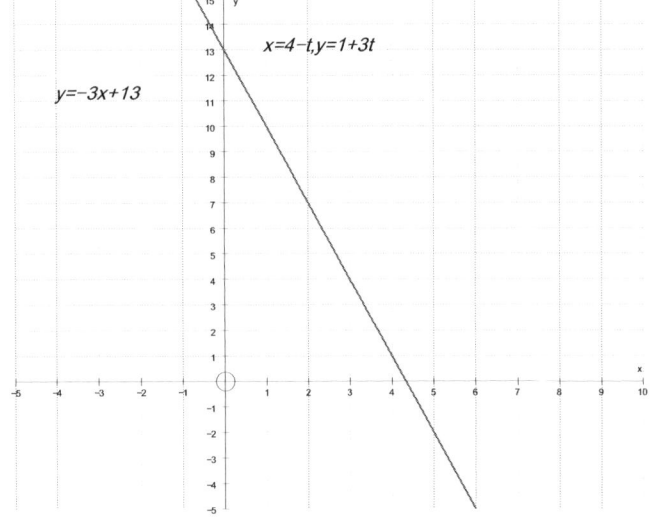

$x = 4 - t$, $y = 1 + 3t$
$\Rightarrow t = 4 - x$
$\Rightarrow y = 1 + 3(4 - x)$
$\Rightarrow y = 1 + 12 - 3x$
$\Rightarrow y = 13 - 3x$
$\Rightarrow y + 3x = 13$

Example

Find the constraint equation of the locus defined by
x= 3+4sinθ
y= 2-5cosθ.

$x = 3 + 4\sin\theta$ $\qquad\qquad y = 2 - 5\cos\theta$

$\Rightarrow \sin\theta = \dfrac{x-3}{4}$ $\qquad\qquad \Rightarrow \cos\theta = \dfrac{y-2}{-5}$

$\Rightarrow \sin^2\theta = \left(\dfrac{x-3}{4}\right)^2$ $\qquad\qquad \Rightarrow \cos^2\theta = \left(\dfrac{y-2}{-5}\right)^2$

$\Rightarrow \sin^2\theta + \cos^2\theta = \left(\dfrac{x-3}{4}\right)^2 + \left(\dfrac{y-2}{-5}\right)^2$

$\Rightarrow 1 = \dfrac{(x-3)^2}{16} + \dfrac{(y-2)^2}{25}$

Which is the equation of an ellipse, major axis parallel to the x axis, centre (3,2), major axis length 8, minor axis length 10.

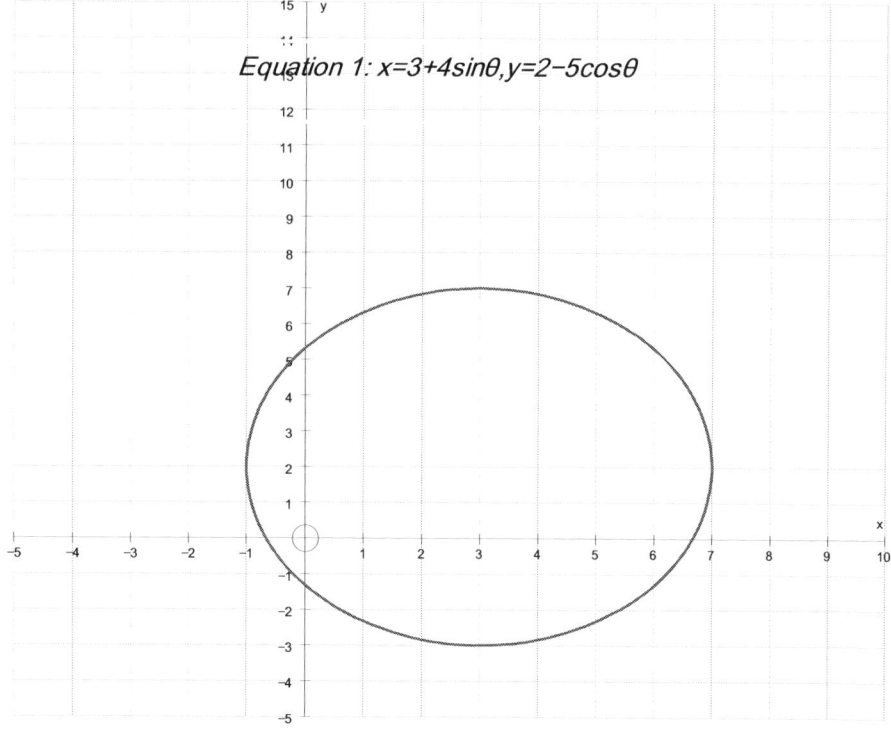

Equation 1: x=3+4sinθ, y=2-5cosθ

DIFFERENTIATING PARAMETRIC EQUATIONS

First derivative.

When x=f(t) and y=g(t) and y=f(x),

then

$$f'x = \frac{dy}{dx}$$
$$= \frac{dy}{dt} \times \frac{dt}{dx}$$
$$= \frac{dy}{dt} \div \frac{dx}{dt}$$

$$\boxed{f'x = \frac{y'(t)}{x'(t)}}$$

Example

Find an expression for the gradient of the curve defined by $x = t^2$, $y = 4t$.

$x = t^2$ $\qquad\qquad$ $y = 4t$
$x'(t) = 2t$ $\qquad\qquad$ $y'(t) = 4$

$$\frac{dy}{dx} = \frac{y'(t)}{x'(t)}$$
$$= \frac{4}{2t}$$
$$= \frac{2}{t}$$

Example

Find the equation of the normal at t=3 on the curve:

$x=t^2$, $y=t^3$

$x = t^2$ $\qquad\qquad$ $y = t^3$

$x'(t) = 2t$ $\qquad\qquad$ $y'(t) = 3t^2$

$$\frac{dy}{dx} = \frac{y'(t)}{x'(t)}$$

$$= \frac{3t^2}{2t}$$

$$= \frac{3t}{2}$$

at t=3, $\quad m = \dfrac{dy}{dx} = \dfrac{9}{2}$

$\therefore$ gradient of normal $= \dfrac{-2}{9}$

$x = 3^2 = 9$ $\qquad\qquad$ $y = 3^3 = 27$

$y - 27 = \dfrac{-2}{9}(x - 9)$

$9y - 243 = -2x + 18$

$9y = -2x + 261$

$y = \dfrac{-2}{9}x + 29$

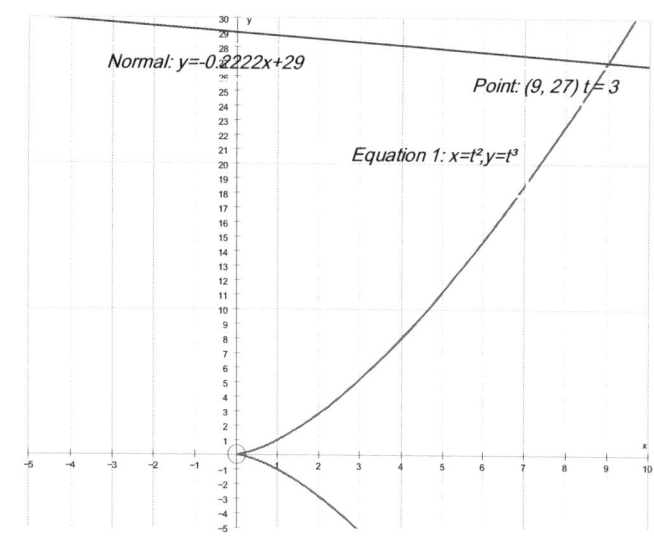

SECOND DERIVATIVE (PARAMETRICS)

$$f''(x) = \frac{d^2y}{dx^2}$$

$$= \frac{d}{dx}\left(\frac{dy}{dx}\right)$$

$$= \frac{d}{dt}\left(\frac{dy}{dx}\right) \cdot \frac{dt}{dx}$$

$$= \frac{d}{dt}\left(\frac{y'(t)}{x'(t)}\right) \cdot \frac{1}{\frac{dx}{dt}} \quad \text{substituting equation for f'(x)}$$

$$= \frac{d}{dt}\left(\frac{y'(t)}{x'(t)}\right) \cdot \frac{1}{x'(t)}$$

$$= \left(\frac{y''(t)x'(t) - x''(t)y'(t)}{(x'(t))^2}\right) \cdot \frac{1}{x'(t)} \quad \text{applying quotient rule}$$

$$= \frac{y''(t)x'(t) - x''(t)y'(t)}{(x'(t))^3}$$

$$\boxed{f''x = \frac{x'(t)y''(t) - x''(t)y'(t)}{(x'(t))^3}}$$

Example

Describe the nature of the curve:

$x=t^2$, $y=t^3$ at the point $t=3$.

$$x = t^2 \qquad\qquad y = t^3$$
$$x'(t) = 2t \qquad\qquad y'(t) = 3t^2$$
$$x''(t) = 2 \qquad\qquad y''(t) = 6t$$

$$\frac{d^2y}{dx^2} = \frac{x'(t)y''(t) - x''(t)y'(t)}{(x'(t))^3}$$

$$= \frac{2t \cdot 6t - 2 \cdot 3t^2}{(2t)^3}$$

$$= \frac{12t^2 - 6t^2}{8t^3}$$

$$= \frac{6t^2}{8t^3}$$

$$= \frac{3}{4t}$$

at t=3, $\quad \dfrac{d^2y}{dx^2} = \dfrac{3}{12}$

$\therefore \dfrac{d^2y}{dx^2} > 0$, so graph is concave up

Example

Find the critical points of the following curve :-

$$x = \frac{3t^2}{1+t^3}, \quad y = \frac{3t}{1+t^3}$$

$$x = \frac{3t^2}{1+t^3}$$

$$x'(t) = \frac{6t(1+t^3) - 3t^2 \cdot 3t^2}{(1+t^3)^2}$$

$$= \frac{6t + 6t^4 - 9t^4}{(1+t^3)^2}$$

$$= \frac{6t - 3t^4}{(1+t^3)^2}$$

$$= \frac{3t(2-t^3)}{(1+t^3)^2}$$

$$y = \frac{3t}{1+t^3}$$

$$y'(t) = \frac{3(1+t^3) - 3t \cdot 3t^2}{(1+t^3)^2}$$

$$= \frac{3 + 3t^3 - 9t^3}{(1+t^3)^2}$$

$$= \frac{3 - 6t^3}{(1+t^3)^2}$$

$$= \frac{3(1-2t^3)}{(1+t^3)^2}$$

$$\frac{dy}{dx} = \frac{y'(t)}{x'(t)}$$

$$= \frac{3(1-2t^3)}{(1+t^3)^2} \times \frac{(1+t^3)^2}{3t(2-t^3)}$$

$$= \frac{3(1-2t^3)}{3t(2-t^3)}$$

$$= \frac{1-2t^3}{2t-t^4}$$

Turning points occur when $\dfrac{dy}{dx} = 0$

$$0 = \frac{1-2t^3}{2t-t^4}$$

$$\Rightarrow 1 - 2t^3 = 0$$

$$\Rightarrow t = \sqrt[3]{\frac{1}{2}}$$

at $t = \sqrt[3]{\dfrac{1}{2}}$,

$x = \dfrac{3t^2}{1+t^3}$ $\qquad$ $y = \dfrac{3t}{1+t^3}$

$x = \dfrac{3\left(\sqrt[3]{\dfrac{1}{2}}\right)^2}{1+\left(\sqrt[3]{\dfrac{1}{2}}\right)^3}$ $\qquad$ $y = \dfrac{3\left(\sqrt[3]{\dfrac{1}{2}}\right)}{1+\left(\sqrt[3]{\dfrac{1}{2}}\right)^3}$

$x = \dfrac{3\left(\dfrac{1}{2}\right)^{\frac{2}{3}}}{1+\dfrac{1}{2}}$ $\qquad$ $y = \dfrac{3\left(\dfrac{1}{2}\right)^{\frac{1}{3}}}{1+\dfrac{1}{2}}$

$x = \dfrac{\left(\dfrac{3}{2^{\frac{2}{3}}}\right)}{\dfrac{3}{2}}$ $\qquad$ $y = \dfrac{\left(\dfrac{3}{2^{\frac{1}{3}}}\right)}{\dfrac{3}{2}}$

$x = \dfrac{6}{3 \times 2^{\frac{2}{3}}}$ $\qquad$ $y = \dfrac{6}{3 \times 2^{\frac{1}{3}}}$

$x = \dfrac{2}{2^{\frac{2}{3}}}$ $\qquad$ $y = \dfrac{2}{2^{\frac{1}{3}}}$

$x = 2^{\frac{1}{3}}$ $\qquad$ $y = 2^{\frac{2}{3}}$

Turning point at $\left(2^{\frac{1}{3}}, 2^{\frac{2}{3}}\right)$

There is a critical point when $t=0$ since $\dfrac{dy}{dx} = \dfrac{1-2t^3}{2t-t^4}$ is undefined this occurs at the point $(0,0)$.

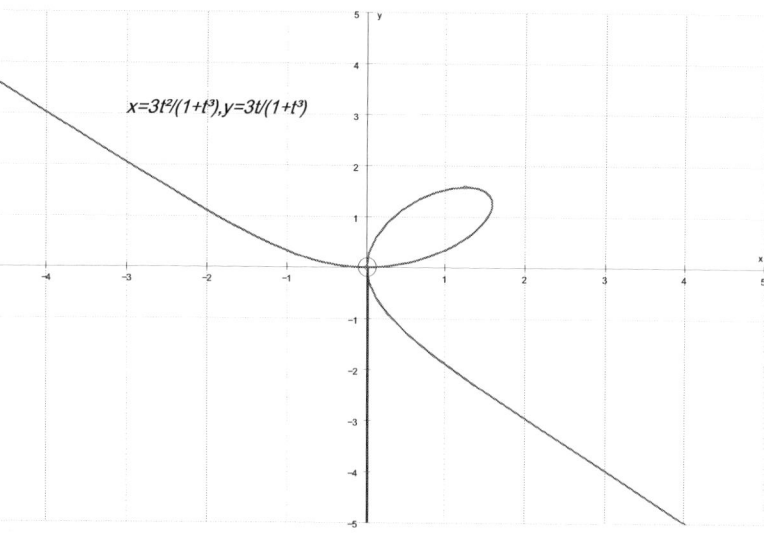

$x=3t^2/(1+t^3), y=3t/(1+t^3)$

OTHER USES

For some y=f(x)

$$f'(x) = \frac{dy}{dx} = \frac{dy}{du} \times \frac{du}{dx}$$

Example

The radius, r cm, of a circular pool of water is increasing at a constant rate of 2cm per minute. Find the rate at which the area of the pool is increasing with time, when the radius is 8cm.

Want: $\dfrac{dA}{dt}$

Know: $\dfrac{dA}{dr} = \dfrac{dA}{dt} \times \dfrac{dt}{dr}$ $\qquad \dfrac{dr}{dt} = 2$ cm/min

Area of a circle = πr^2

$$\Rightarrow \frac{dA}{dr} = 2\pi r$$

$$\frac{dA}{dr} = \frac{dA}{dt} \times \frac{dt}{dr}$$

$$\Rightarrow 2\pi r = \frac{dA}{dt} \times \frac{1}{2}$$

$$\Rightarrow 4\pi r = \frac{dA}{dt}$$

when r = 8

$$\frac{dA}{dt} = 32\pi \text{ cm/min}$$

INTEGRATION

*MAC 1.3: **Applying calculus skills through techniques of integration.***

$$\int f(x)dx = F(x)+c$$

Integral calculus is mainly concerned with summing the values of a function over a particular range, and is particularly useful for finding the area of an irregular shape.

Integration is the process of finding the anti-derivative.

$\int f(x)dx$ means the indefinite integral of the function is to be calculated with respect to x. The anti-derivative F(x) is called the integral, c is called the constant of integration.

$F'(x) = f(x)$ so differentiating the integral results in the original function.

Basic Rules of Integration

$$\int x^n dx = \frac{x^{n+1}}{n+1} + c, \quad n \neq -1$$

$$\int (ax+b)^n dx = \frac{(ax+b)^{n+1}}{a(n+1)} + c, \quad n \neq -1$$

Constant Multiplier Rule

$$\int kf(x)dx = k\int f(x)dx, \quad k \text{ is a constant}$$

Sum Rule

$$\int (f(x) + g(x))dx = \int f(x)dx + \int g(x)dx$$

Examples

$\int x\,dx = \frac{x^2}{2} + c$

$\int \sqrt{x}\,dx = \int x^{\frac{1}{2}} dx$

$= \frac{x^{\frac{3}{2}}}{\frac{3}{2}} + c$

$= \frac{2}{3} x^{\frac{3}{2}} + c$

$\int \sqrt[5]{x}\,dx = \int x^{\frac{1}{5}} dx = \frac{5}{6} x^{\frac{6}{5}} + c$

$\int (5t^3 - 8t)dt = 5\int t^3 dt - 8\int t\,dt$

$= 5\frac{t^4}{4} - 8\frac{t^2}{2} + c$

$= \frac{5t^4}{4} - 4t^2 + c$

$= 4t^2 \left(\frac{5t^2}{16} - 1\right) + c$

$\int (4x^2 - 2x)dx = \int 4x^2 dx - \int 2x\,dx$

$= 4\int x^2 dx - 2\int x\,dx$

$= 4\frac{x^3}{3} - 2\frac{x^2}{2} + c$

$= \frac{4x^3}{3} - x^2 + c$

$= x^2 \left(\frac{4x}{3} - 1\right) + c$

Never forget the c !

TRIG FUNCTIONS

Remember to use Radians!!!

$\int \cos x \, dx = \sin x + c$

$\int \sin x \, dx = -\cos x + c$

$\int \cos(ax+b) \, dx = \frac{1}{a}\sin(ax+b) + c$

$\int \sin(ax+b) \, dx = -\frac{1}{a}\cos(ax+b) + c$

$\int \sec^2 x \, dx = \tan x + c$

Examples

$\int 5\cos x \, dx = 5\sin x + c$

$\int 10\sin x \, dx = -10\cos x + c$

$\int \cos(5x+6) \, dx = \frac{1}{5}\sin(5x+6) + c$

$\int \sin(3x-2) \, dx = -\frac{1}{3}\cos(3x-2) + c$

$\int (3x + \cos(5x+6)) \, dx = \int 3x \, dx + \int \cos(5x+6) \, dx$

$= 3\int x \, dx + \int \cos(5x+6) \, dx$

$= \frac{3x^2}{2} + \frac{1}{5}\sin(5x+6) + c$

$\int_0^{\frac{\pi}{6}} \frac{1}{\cos^2 x} \, dx = \int_0^{\frac{\pi}{6}} \sec^2 x \, dx$

$= [\tan x]_0^{\frac{\pi}{6}}$

$= \tan \frac{\pi}{6} - \tan 0$

$= \frac{1}{\sqrt{3}}$

LOGARITHMS AND EXPONENTIALS

Don't forget to apply the chain rule in reverse if necessary!

$$\int \frac{1}{x}\,dx = \log_e x + C \qquad \int e^x\,dx = e^x + c$$

Examples

$$\int -2e^{-\frac{1}{2}x}\,dx$$
$$= -2\int e^{-\frac{1}{2}x}\,dx$$
$$= -2 \cdot e^{-\frac{1}{2}x} \cdot -2 + C$$
$$= 4e^{-\frac{1}{2}x} + C$$

$$\int \frac{1}{3x}\,dx = \frac{1}{3}\int \frac{1}{x}\,dx$$
$$= \frac{1}{3}\log_e |x| + C$$
$$= \ln x^{\frac{1}{3}} + C$$

$$\int_0^1 \frac{(e^x - 1)^2}{e^{2x}}\,dx = \int_0^1 \frac{e^{2x} - 2e^x + 1}{e^{2x}}\,dx$$

$$= \int_0^1 1 - 2e^{-x} + e^{-2x}\,dx$$

$$= \left[x - \frac{2e^{-x}}{-1} + \frac{e^{-2x}}{-2} \right]_0^1$$

$$= \left(1 + 2e^{-1} - \frac{e^{-2}}{2}\right) - \left(0 + 2e^0 - \frac{e^0}{2}\right)$$

$$= 1 + \frac{2}{e} - \frac{1}{2e^2} - 0 - 2 + \frac{1}{2}$$

$$= \frac{2}{e} - \frac{1}{2e^2} - \frac{1}{2}$$

INTEGRATION BY SUBSTITUTION

When faced with an integral of the form
$\int g(f(x)) \cdot f'(x) dx$
proceed as follows :-

1) Identify the essential function, f(x)
2) Let u = f(x), so du = f'(x)dx
3) Substitute these values into the original integral
4) Integrate
5) substitute back

Examples

Find $\int 5x^2(x^3 - 4)^4 dx$ let $u = x^3 - 4$

then $du = 3x^2 dx$

$= \int \frac{5}{3} U^4 du$

$= \frac{5}{3} U^5 \cdot \frac{1}{5} + C$

$= \frac{1}{3} U^5 + C$

$= \frac{1}{3}(x^3 - 4)^5 + C$

Find $\int e^x \sin(e^x) dx$ let $u = e^x$

then $du = e^x dx$

$= \int \sin u \, du$

$= -\cos u$

$= -\cos e^x + C$

Find $\int_0^{\frac{1}{2}} \frac{x}{\sqrt{1-x^2}} dx$

let $u = 1 - x^2$
then $du = -2x dx$
when $x = \frac{1}{2}$, $u = 1 - (\frac{1}{2})^2 = \frac{3}{4}$
when $x = 0$, $u = 1$

$= -\frac{1}{2} \int_1^{\frac{3}{4}} \frac{1}{\sqrt{u}} du$

$= -\frac{1}{2} \left[2u^{\frac{1}{2}} \right]_1^{\frac{3}{4}}$

$= \left[-u^{\frac{1}{2}} \right]_1^{\frac{3}{4}}$

$= -\left(\frac{3}{4}\right)^{\frac{1}{2}} - (-1)$

$= -\frac{\sqrt{3}}{2} + 1$

$= 1 - \frac{\sqrt{3}}{2}$

Find $\int_0^{\frac{\pi}{2}} \sin^5 x \, dx$

let $u = \cos x$
then $du = -\sin x \, dx$
when $x = \frac{\pi}{2}$, $u = \cos(\frac{\pi}{2}) = 0$
when $x = 0$, $u = \cos(0) = 1$

$= \int_0^{\frac{\pi}{2}} \left(\sin^2 x\right)^2 \sin x \, dx$

$= \int_0^{\frac{\pi}{2}} \left(1 - \cos^2 x\right)^2 \sin x \, dx$

$= -\int_1^0 \left(1 - u^2\right)^2 du$

$= -\int_1^0 \left(1 - 2u^2 + u^4\right) du$

$= -\left[u - \frac{2}{3} u^3 + \frac{1}{5} u^5 \right]_1^0$

$= \left[-u + \frac{2}{3} u^3 - \frac{1}{5} u^5 \right]_1^0$

$= 0 - \left(-1 + \frac{2}{3} - \frac{1}{5} \right)$

$= 1 - \frac{2}{3} + \frac{1}{5}$

$= \frac{8}{15}$

Sometimes it is easier to substitute dx

Find $\int x \sqrt[4]{5+x}\, dx$

let $u^4 = 5 + x$

$u = \sqrt[4]{5+x}$

and $x = u^4 - 5$

so $dx = 4u^3 du$

$= \int (u^4 - 5) \cdot u \cdot 4u^3 du$

$= \int (u^4 - 5) 4u^4 du$

$= 4\int u^8 - 5u^4 du$

$= 4\left(\dfrac{u^9}{9} - u^5\right) + c$

$= \dfrac{4u^9}{9} - 4u^5 + c$

$= \dfrac{4\left(\sqrt[4]{5+x}\right)^9}{9} - 4\left(\sqrt[4]{5+x}\right)^5 + c$

$= \dfrac{4(5+x)^{9/4}}{9} - 4(5+x)^{5/4} + c$

Find $\int \dfrac{x^5 dx}{(1+x^2)^{3/2}}$

let $x = \sqrt{u^2 - 1}$

then $dx = \dfrac{1}{2}(u^2-1)^{-\frac{1}{2}} \cdot 2u\, du$

so $dx = u(u^2 - 1)^{-\frac{1}{2}} du$

and $x^2 = u^2 - 1$

so $\sqrt{x^2 + 1} = u$

$= \int \dfrac{\left(\sqrt{u^2-1}\right)^5 \cdot u(u^2-1)^{-\frac{1}{2}} du}{(u^2)^{3/2}}$

$= \int \dfrac{(u^2-1)^{5/2} \cdot u(u^2-1)^{-\frac{1}{2}} du}{u^3}$

$= \int \dfrac{u(u^2-1)^2 du}{u^3}$

$= \int \dfrac{u(u^4 - 2u^2 + 1) du}{u^3}$

$= \int \dfrac{u^5 - 2u^3 + u}{u^3} du$

$= \int u^2 - 2 + u^{-2}\, du$

$= \dfrac{u^3}{3} - 2u - u^{-1} + c$

$= \dfrac{\left(\sqrt{x^2+1}\right)^3}{3} - 2\left(\sqrt{x^2+1}\right) - \left(\sqrt{x^2+1}\right)^{-1} + c$

$= \dfrac{\left(\sqrt{x^2+1}\right)^3}{3} - 2\sqrt{x^2+1} - \dfrac{1}{\sqrt{x^2+1}} + c$

COMMON FORMS

$$\int f(ax+b)dx = \frac{1}{a}F(ax+b)+C$$

Examples

$$\int (4x+2)^4 dx = \frac{1}{4} \cdot \frac{1}{5}(4x+2)^5 + C$$
$$= \frac{1}{20}(4x+2)^5 + C$$

Proof

$$\int (4x+2)^4 dx \qquad \text{let } u = 4x+2$$
$$\text{then } du = 4\,dx$$
$$\Rightarrow dx = \frac{du}{4}$$

$$= \int u^4 \cdot \frac{1}{4} du$$
$$= \frac{1}{4}\int u^4 du$$
$$= \frac{1}{4} \cdot \frac{1}{5}u^5 + C$$
$$= \frac{1}{20}u^5 + C$$
$$= \frac{1}{20}(4x+2)^5 + C$$

$$\int \cos(3-5x)dx = \frac{1}{-5}\sin(3-5x)+C$$

Proof

$$\int \cos(3-5x)dx \qquad \text{let } u = 3-5x$$
$$\text{then } du = -5\, dx$$
$$\Rightarrow dx = \frac{du}{-5}$$

$$= \int \cos u \cdot \frac{1}{-5}du$$

$$= \frac{1}{-5}\int \cos u\, du$$

$$= \frac{1}{-5}\sin u + C$$

$$= -\frac{1}{5}\sin(3-5x)+C$$

$$\int e^{(4+8x)}dx = \frac{1}{8}e^{(4+8x)}+C$$

Proof

$$\int e^{(4+8x)}dx \qquad \text{let } u = 4+8x$$
$$\text{then } du = 8\, dx$$
$$\Rightarrow dx = \frac{du}{8}$$

$$= \int e^u \cdot \frac{1}{8}du$$

$$= \frac{1}{8}\int e^u\, du$$

$$= \frac{1}{8}e^u + C$$

$$= \frac{1}{8}e^{(4+8x)}+C$$

$$\boxed{\int \frac{f'(x)}{f(x)} dx = \ln|f(x)| + C}$$

Examples

$$\int \frac{4x^3}{7+x^4} dx = \ln|7+x^4| + C$$

Proof

$$\int \frac{4x^3}{7+x^4} dx \qquad\qquad let\ u = 7+x^4$$
$$\qquad\qquad then\ du = 4x^3\ dx$$

$$= \int \frac{du}{u}$$
$$= \ln|u| + c$$
$$= \ln|7+x^4| + c$$

$$\int \frac{6x^2}{8x^3+5} dx = \frac{1}{4}\int \frac{24x^2}{8x^3+5} dx = \frac{1}{4}\ln|8x^3+5| + C$$

Proof

$$\int \frac{6x^2}{8x^3+5} dx \qquad\qquad let\ u = 8x^3+5$$
$$\qquad\qquad then\ du = 24x^2\ dx$$

$$= \frac{1}{4}\int \frac{du}{u}$$
$$= \frac{1}{4}\ln|u| + c$$
$$= \frac{1}{4}\ln|8x^3+5| + C$$

$\int 5\cot x\,dx = 5\int \dfrac{\cos x}{\sin x}\,dx = 5\ln|\sin x| + C$

Proof

$\int 5\cot x\,dx$ let $u = \sin x$

                                                           then du = cosx dx

$= 5\int \dfrac{\cos x\,dx}{\sin x}$

$= 5\int \dfrac{du}{u}$

$= 5\ln|u| + c$

$= 5\ln|\sin x| + C$

$\int 3\cot 2x\,dx = 3\int \dfrac{\cos 2x}{\sin 2x}\,dx = \dfrac{3}{2}\ln|\sin 2x| + C$

Proof

$\int 3\cot 2x\,dx$ let $u = \sin 2x$

                                            then du = 2cos2x dx

$= \dfrac{3}{2}\int \dfrac{2\cos 2x\,dx}{\sin 2x}$

$= \dfrac{3}{2}\int \dfrac{du}{u}$

$= \dfrac{3}{2}\ln|u| + c$

$= \dfrac{3}{2}\ln|\sin 2x| + C$

$$\boxed{\int f'(x)f(x)dx = \frac{1}{2}(f(x))^2 + C}$$

Examples

$$\int (4x+5)(2x^2+5x)dx = \frac{1}{2}(2x^2+5x)^2 + C$$

Proof

$\int (4x+5)(2x^2+5x)dx$ let $u = 2x^2 + 5x$

                                 then $du = (4x+5)dx$

$= \int u\, du$

$= \dfrac{u^2}{2} + C$

$= \dfrac{(2x^2+5x)^2}{2}$

$= \dfrac{1}{2}(2x^2+5x)^2 + C$

$$\boxed{\int x^2(2x^3+5)dx = \frac{1}{6}\int 6x^2(2x^3+5)dx}$$

$$= \frac{1}{6} \cdot \frac{1}{2}(2x^3+5)^2 + C$$

$$= \frac{1}{12}(2x^3+5)^2 + C$$

Proof

$\int x^2(2x^3+5)dx$ let $u = 2x^3 + 5$

                          then $du = 6x^2 dx$

$= \dfrac{1}{6}\int u\, du$

$= \dfrac{1}{6} \cdot \dfrac{u^2}{2} + C$

$= \dfrac{1}{6} \cdot \dfrac{(2x^3+5)^2}{2}$

$= \dfrac{1}{12}(2x^3+5)^2 + C$

INFINITE INTEGRALS

These occur when the limits of integration or the integrand become infinite.

Limits of integration become infinite:-

let $F'(x) = f(x)$ for all values of $x > a$,
where a is a finite real number

$$\int_a^\infty f(x)\,dx = \lim_{X \to \infty} \int_a^X f(x)\,dx$$

$$= \lim_{X \to \infty} \int_a^X F(x) - F(a) \quad \text{when this limit exists}$$

$$\int_{-\infty}^b f(x)\,dx = \lim_{X \to -\infty} \int_a^X f(x)\,dx$$

$$= F(b) - \lim_{X \to -\infty} \int_X^b F(x) \quad \text{when this limit exists}$$

If the limit exists the integral is convergent.

Otherwise, it is divergent.

Examples

is $\quad \int_1^\infty \dfrac{3}{x} \, dx \quad$ convergent?

$$\int_1^\infty \dfrac{3}{x} \, dx = \lim_{X \to \infty} \int_1^X \dfrac{3}{x} \, dx$$
$$= \lim_{X \to \infty} \left[3 \ln x \right]_1^X$$
$$= \lim_{X \to \infty} \left[3 \ln X - 3 \ln 1 \right]$$
$$= \infty$$

this limit does not exist, so the integral is divergent.

Evaluate $\quad \int_1^\infty e^{-kx} \, dx \, , \, k > 0$

$$\int_0^\infty e^{-kx} \, dx = \lim_{X \to \infty} \int_0^X e^{-kx} \, dx$$
$$= \lim_{X \to \infty} \left[e^{-kx} \cdot \dfrac{-1}{k} \right]_0^X$$
$$= \lim_{X \to \infty} \left[\dfrac{-e^{-kx}}{k} \right]_0^X$$
$$= \lim_{X \to \infty} \left(\dfrac{-e^{-kX}}{k} + \dfrac{e^0}{k} \right)$$
$$= \lim_{X \to \infty} \left(\dfrac{1}{k}(1 - e^{-kX}) \right)$$
$$= \dfrac{1}{k}$$

THE INTEGRAND BECOMES INFINITE:-

If the integrand $f(x)$ becomes infinite at the endpoints of the interval $a < x < b$, then the integral does not exist. However, the *improper* integral may exist.

let $F'(x) = f(x)$ for $a < x < b$

If both $F(a+0)$ and $F(b-0)$ tend to a finite limit then

$$\int_a^b f(x)\,dx = F(b-0) - F(a+0)$$

Examples
Infinite Endpoint

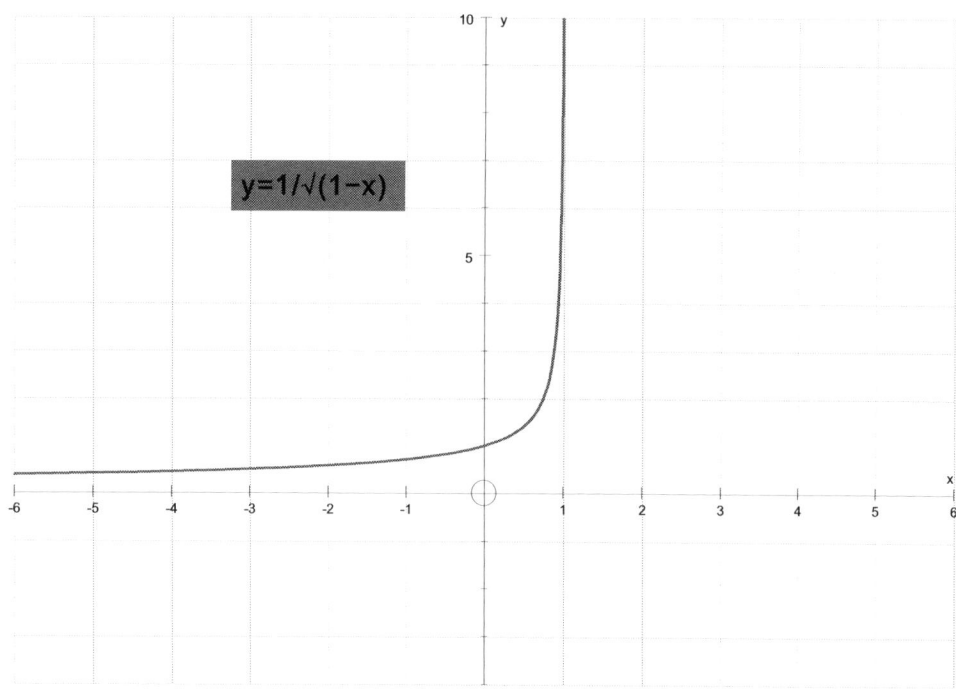

Evaluate $\int_0^1 \frac{1}{\sqrt{(1-x)}} dx$

consider

$$\int_0^{1-\varepsilon} \frac{1}{\sqrt{(1-x)}} dx = \lim_{\varepsilon \to 0} \int_0^{1-\varepsilon} \frac{1}{\sqrt{(1-x)}} dx$$

$$= \lim_{\varepsilon \to 0} \int_0^{1-\varepsilon} (1-x)^{\frac{-1}{2}} dx$$

$$= \lim_{\varepsilon \to 0} \left[-2(1-x)^{\frac{1}{2}} \right]_0^{1-\varepsilon}$$

$$= \lim_{\varepsilon \to 0} \left(-2(1-(1-\varepsilon)^{\frac{1}{2}} + 2 \right)$$

$$= 2$$

The limit exists,

$$\int_0^1 \frac{1}{\sqrt{(1-x)}} dx = 2$$

INFINITE POINT WITHIN THE INTEGRAND.

Example

Evaluate $\int_{-3}^{-1} \dfrac{2}{(x+2)^2}\, dx$

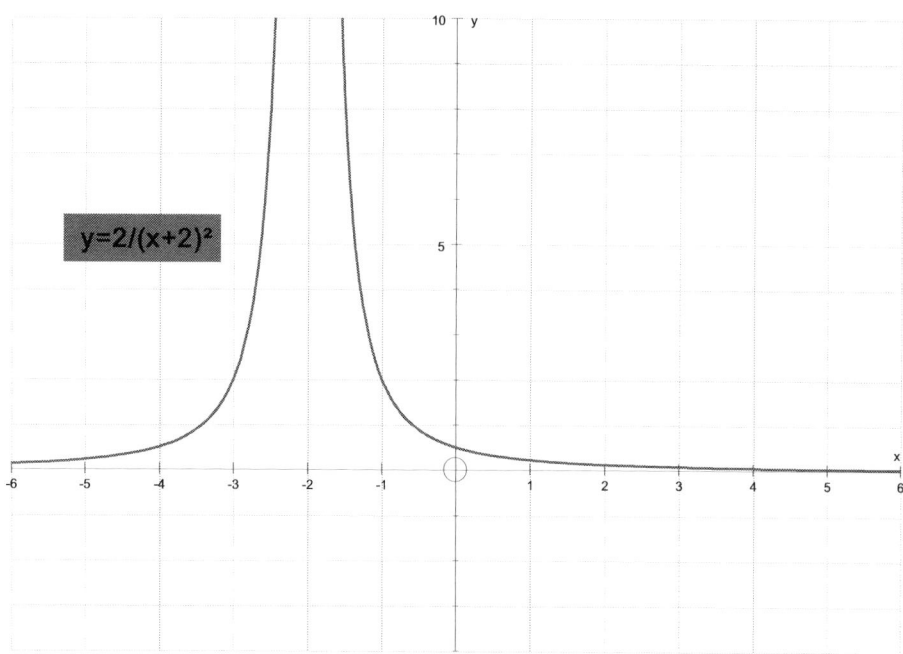

Clearly, the function is undefined at x=-2
The interval must be split and inspected.

Evaluate $\int_{-3}^{-1} \dfrac{2}{(x+2)^2}\,dx$

consider

$$\int_{-2+\varepsilon}^{-1} \dfrac{2}{(x+2)^2}\,dx = \lim_{\varepsilon \to 0}\int_{-2+\varepsilon}^{-1} \dfrac{2}{(x+2)^2}\,dx$$

$$= \lim_{\varepsilon \to 0} 2\int_{-2+\varepsilon}^{-1} (x+2)^{-2}\,dx$$

$$= \lim_{\varepsilon \to 0} \left(-2\left[(x+2)^{-1}\right]_{-2+\varepsilon}^{-1}\right)$$

$$= \lim_{\varepsilon \to 0}\left(-2\left(\dfrac{1}{(-1+2)} - \dfrac{1}{(-2+\varepsilon+2)}\right)\right)$$

$$= \lim_{\varepsilon \to 0}\left(-2\left(1 - \dfrac{1}{\varepsilon}\right)\right)$$

$$\dfrac{1}{\varepsilon} \to \infty \quad \text{as} \quad \varepsilon \to 0$$

The right hand limit does not exist

consider

$$\int_{-3}^{-2-\varepsilon} \dfrac{2}{(x+2)^2}\,dx = \int_{-3}^{-2-\varepsilon} \dfrac{2}{(x+2)^2}\,dx$$

$$= \lim_{\varepsilon \to 0} 2\int_{-3}^{-2-\varepsilon} (x+2)^{-2}\,dx$$

$$= \lim_{\varepsilon \to 0}\left(-2\left[(x+2)^{-1}\right]_{-3}^{-2-\varepsilon}\right)$$

$$= \lim_{\varepsilon \to 0}\left(-2\left(\dfrac{1}{(-2-\varepsilon+2)} - \dfrac{1}{(-3+2)}\right)\right)$$

$$= \lim_{\varepsilon \to 0}\left(-2\left(\dfrac{1}{-\varepsilon} + 1\right)\right)$$

$$\dfrac{1}{\varepsilon} \to \infty \quad \text{as} \quad \varepsilon \to 0$$

The left hand limit does not exist.

The area is unbounded.

The integral is divergent.

AREAS UNDER CURVES

In the following graph, a vehicle starts at rest and accelerates.

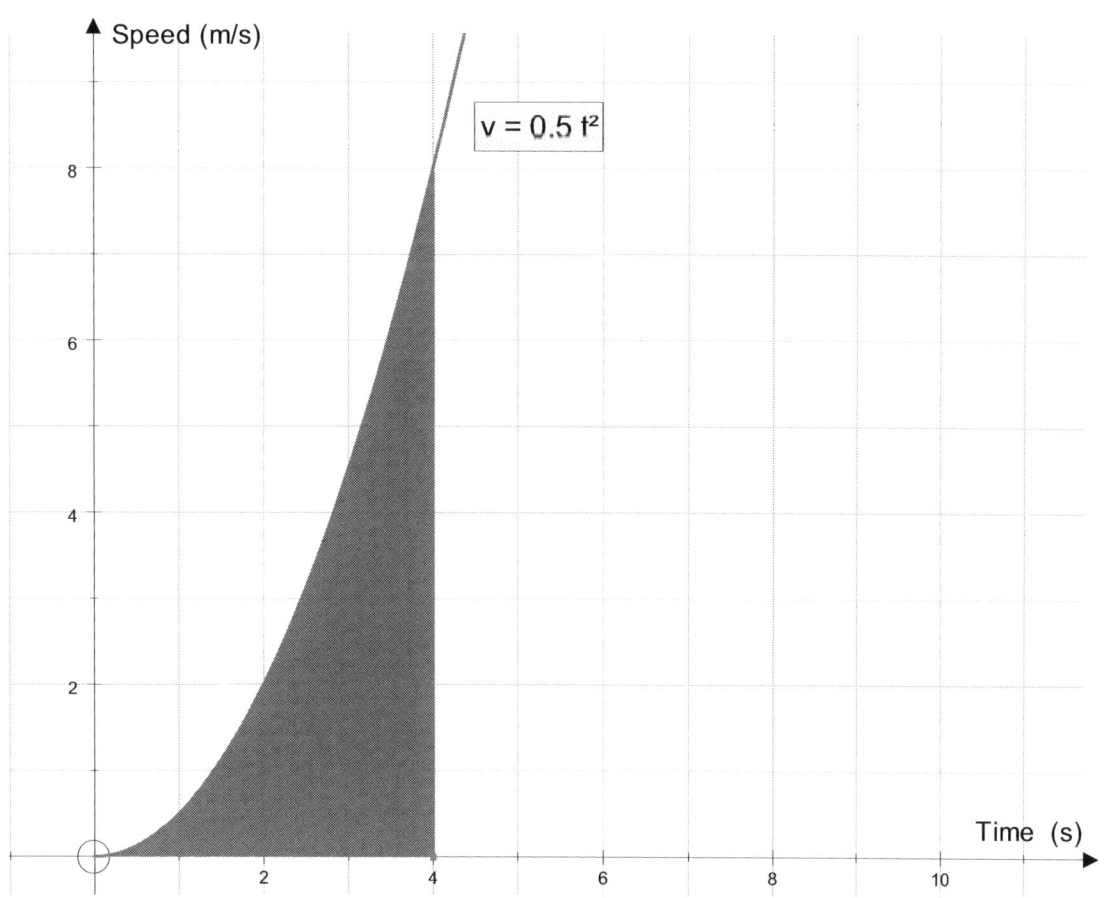

What distance is travelled by the vehicle during the first 4 seconds?

This is more of a challenge.

Split the curve into intervals of ½ seconds and draw rectangles to find upper and lower bounds of the area.

$v = 0.5t^2$

Time (s)	t	0.5	1	1.5	2	2.5	3	3.5	4
Speed (m/s)	v	0.125	0.5	1.125	2	3.125	4.5	6.125	8

The **lower bound** of the area under the curve is the sum of the areas of the rectangles drawn to the right of the curve.

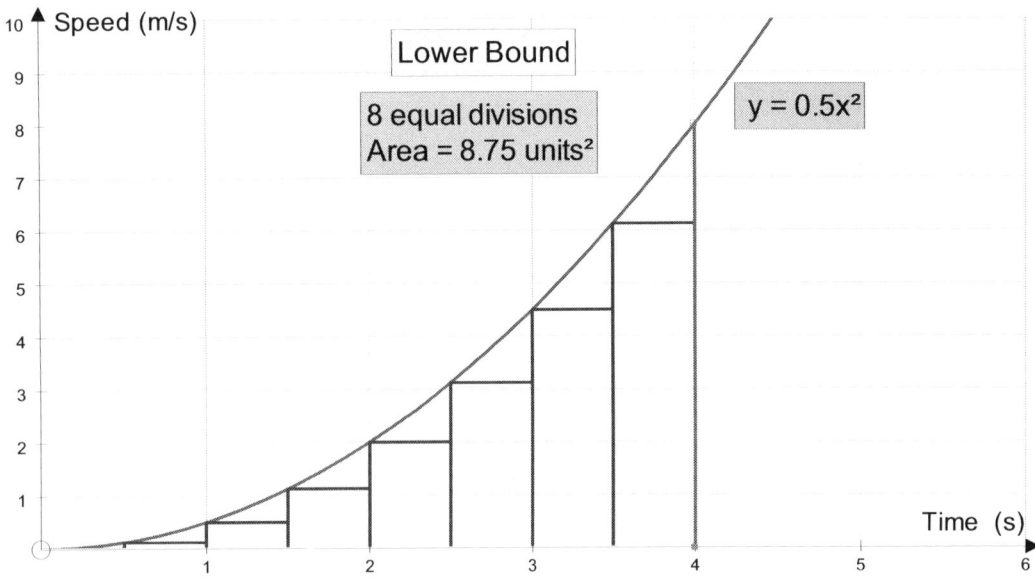

Area = 0.5(0+0.125+0.5+1.125+2+3.125+4.5+6.125)
Area = 0.5 x 17.5 = 8.75 units²

The **upper bound** of the area under the curve is the sum of the areas of the rectangles drawn to the left of the curve.

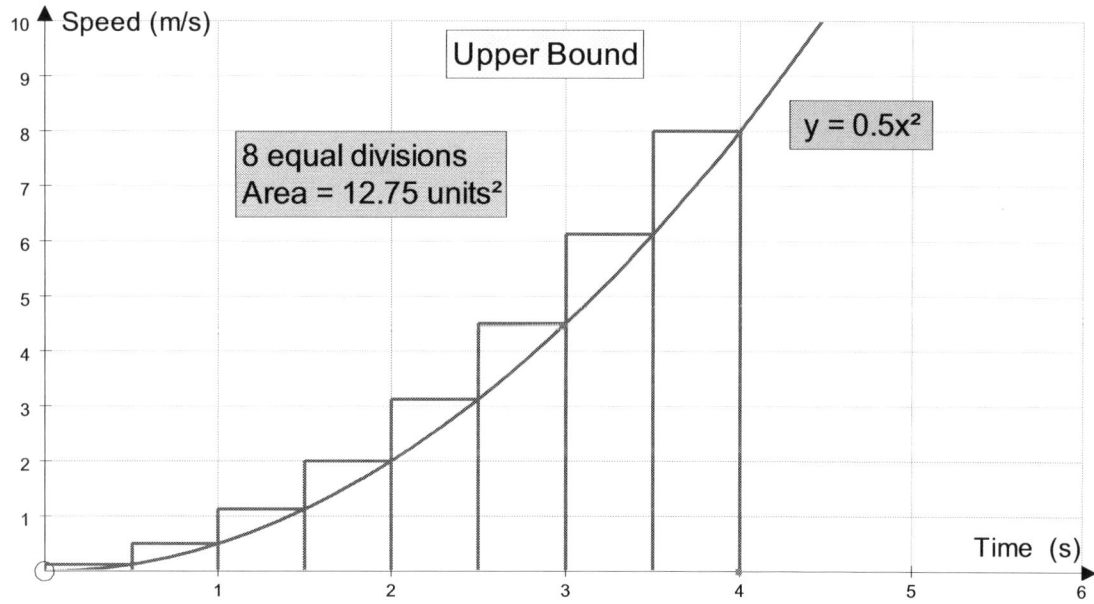

Area = 0.5(0.125+0.5+1.125+2+3.125+4.5+6.125+8)
Area = 0.5 x 25.5 = 12.75 units²

The true area under the graph lies somewhere between these two values.

Increasing the number of rectangles helps:-

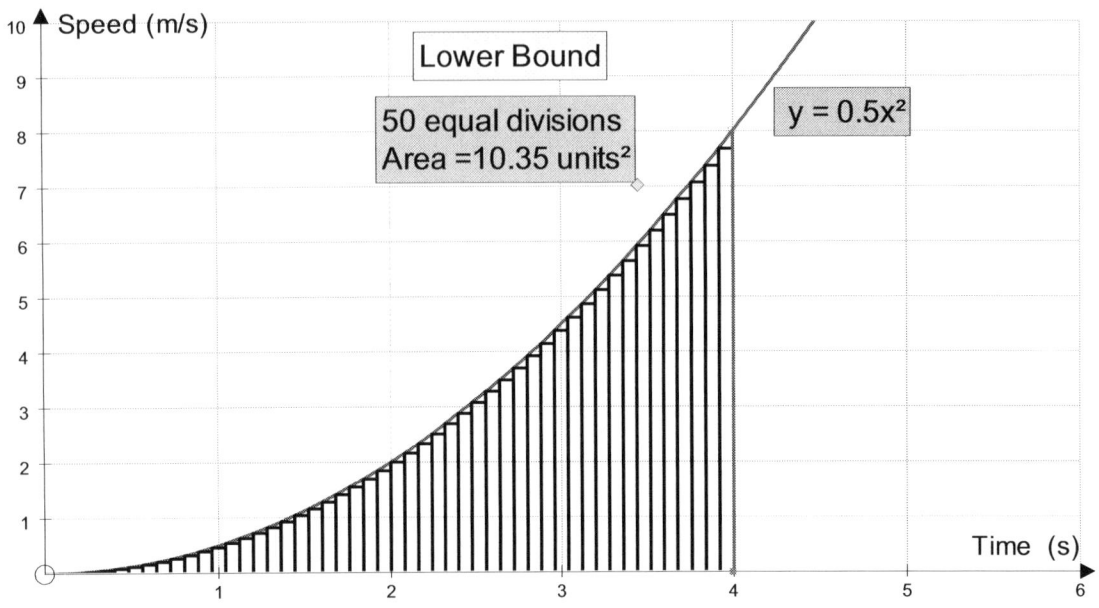

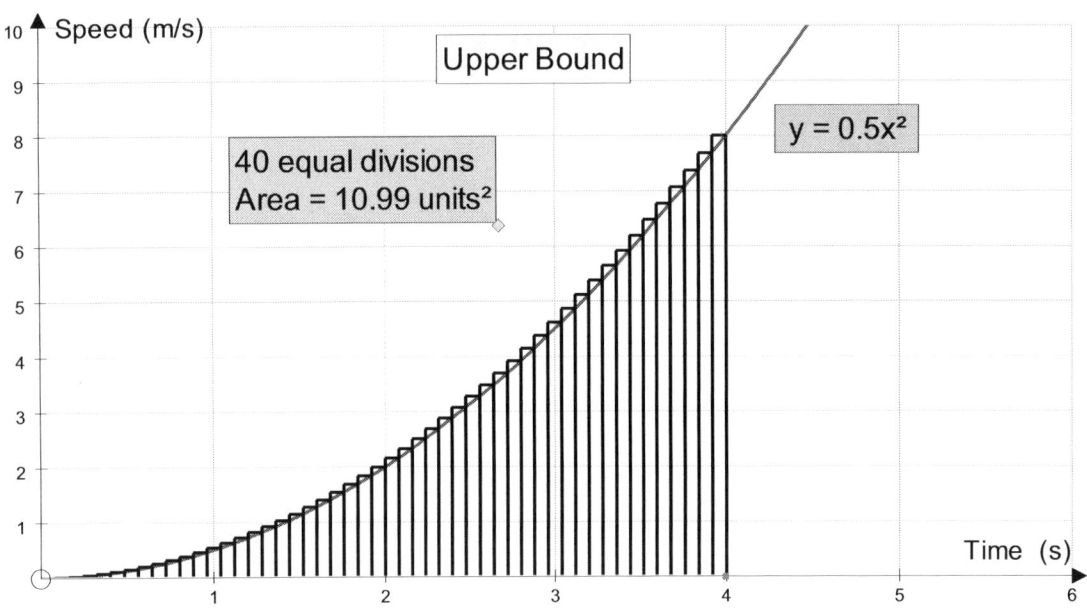

Eventually, both the upper and lower bounds converge to a limit.

This limit is the area under the graph.

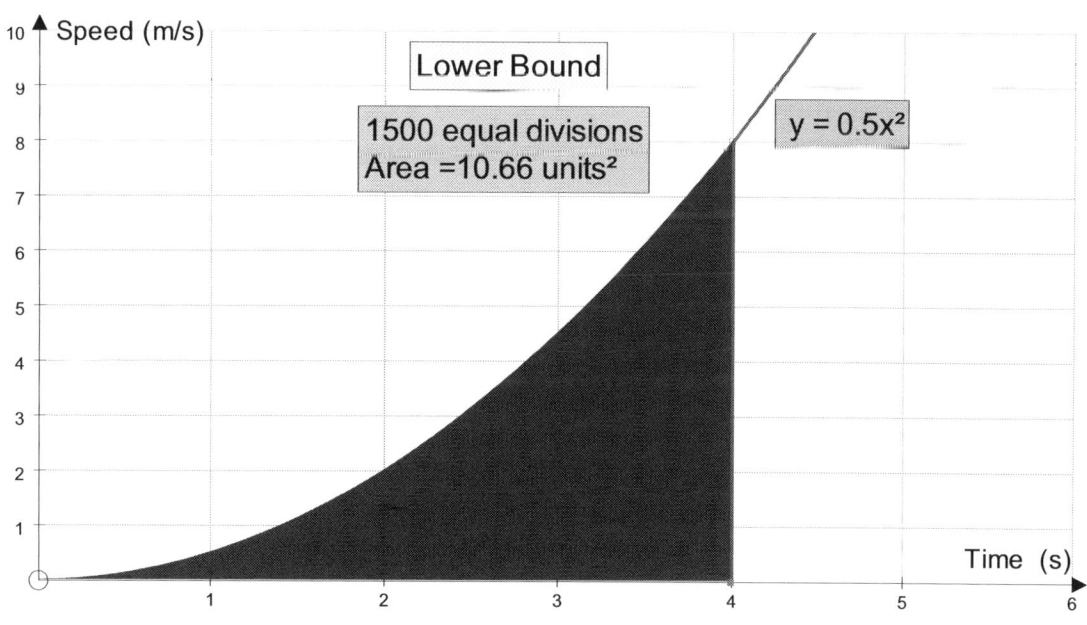

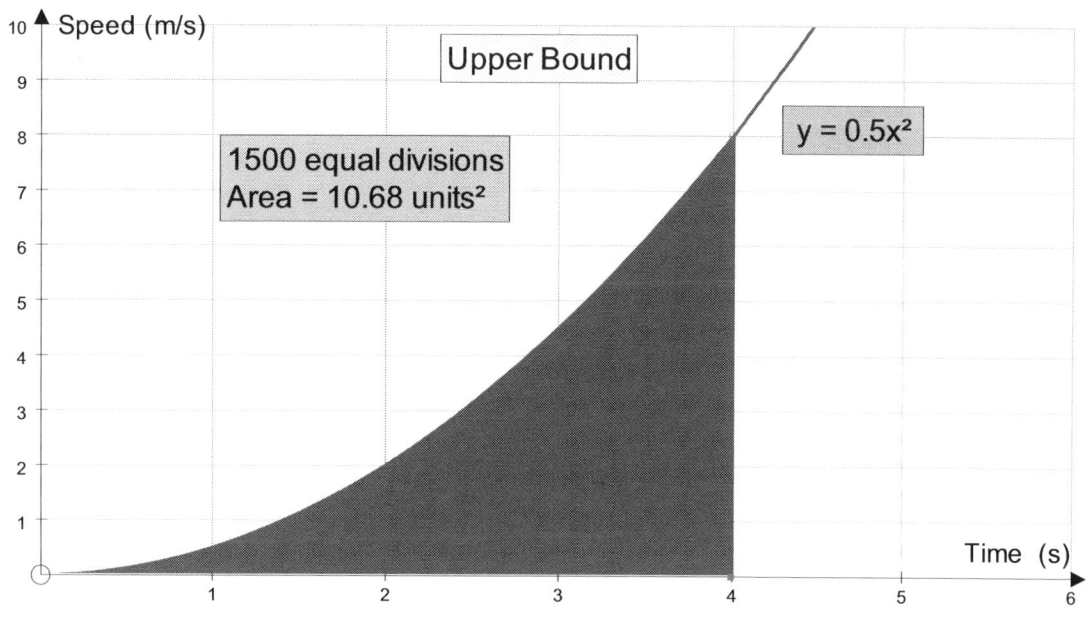

Here, the area is converging to 10.67 units²

INTEGRATION AND THE AREA FUNCTION

The area between the graph of the function y = f(x) and the x-axis, starting at x = 0 is called the area function A(x).

When $y = (n+1)x^n$, $A(x) = x^{(n+1)}$, $(n \neq -1)$

Example
Find the area under the graph y = 2x between x = 2 and x = 4

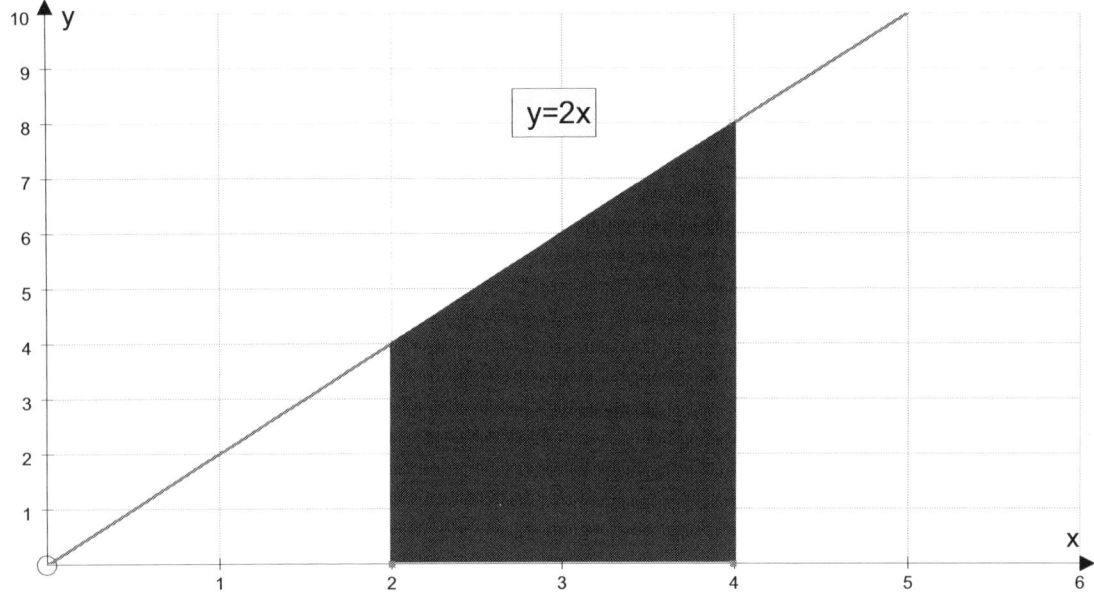

The area between 2 and 4 can be described as e area between x = 0 and x = 4 minus the area between x = 0 and x = 2

y = 2x so $A(x) = x^2$
$A(4) - A(2) = 4^2 - 2^2 = 16 - 4 = 12$ units²

But $\int 2x\,dx = x^2$ so $A(x) = \int f(x)\,dx$

DEFINITE INTEGRALS

The area of the graph of y = f(x) between x = a and x = b is

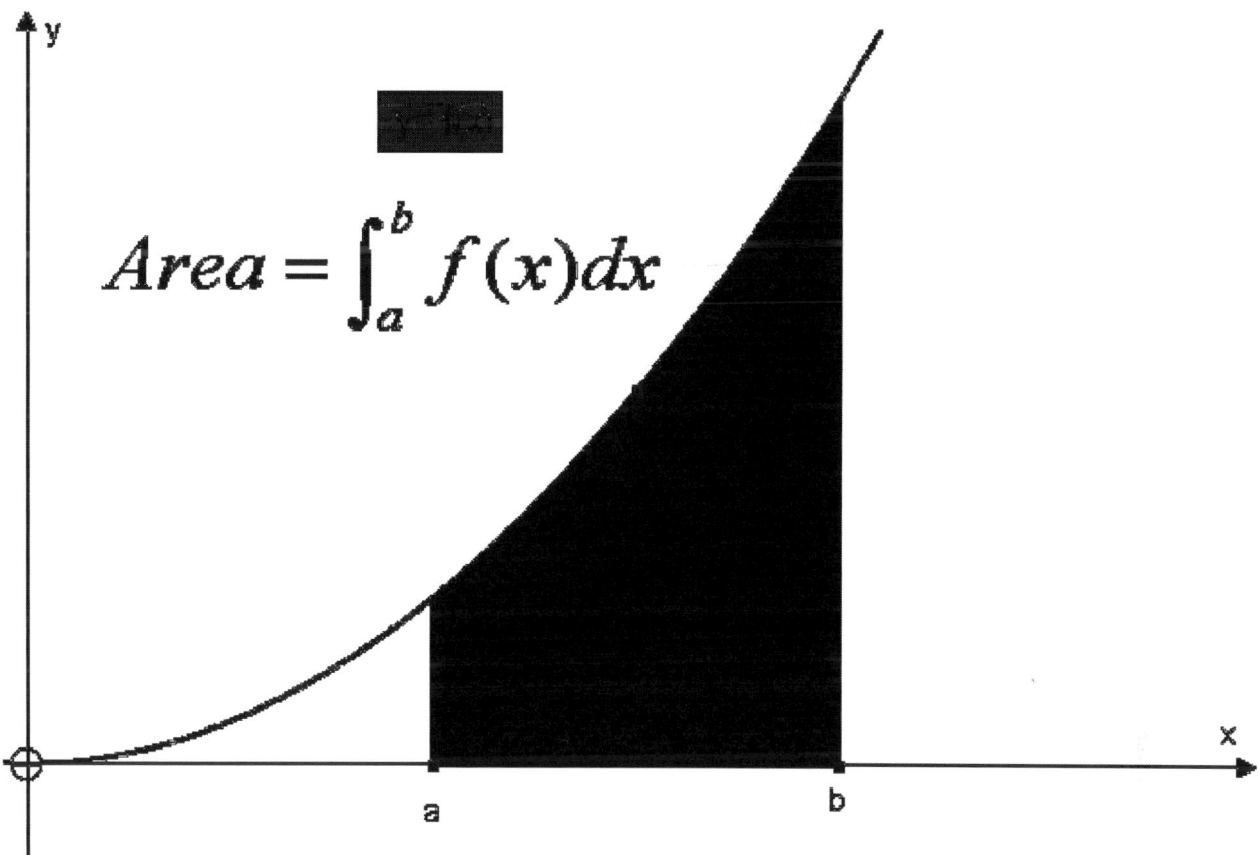

$\int_a^b f(x)dx$ is called a definite integral.

b is the upper limit of integration

a is the lower limit of integration

Example

Find the shaded area as a definite integral.

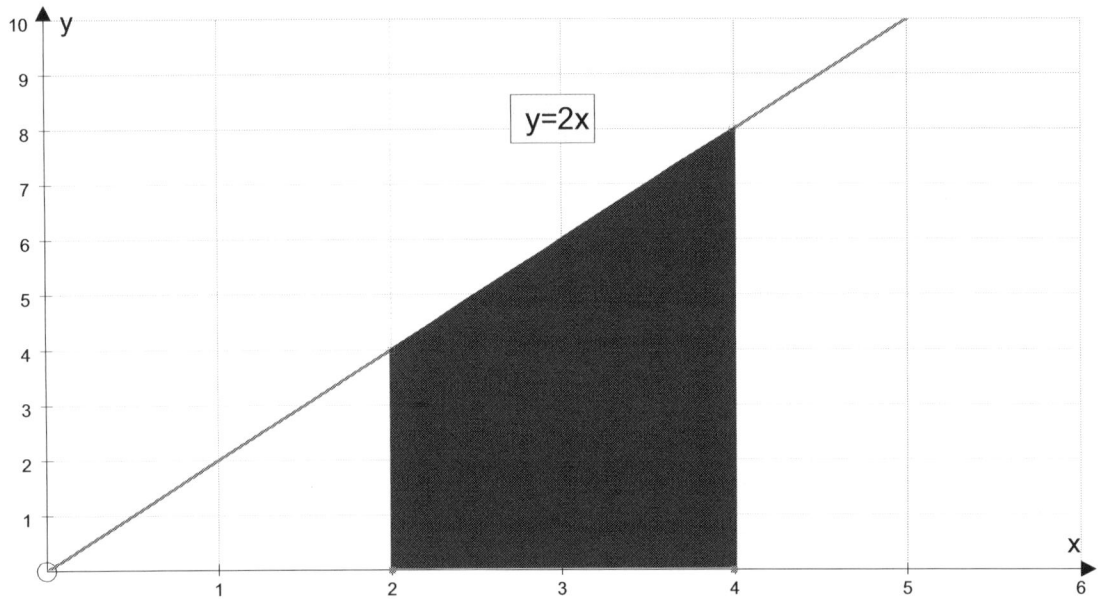

$$Area = \int_2^4 2x\,dx$$
$$= \left[x^2\right]_2^4$$
$$= (4^2) - (2^2)$$
$$= 16 - 4$$
$$= 12 \text{ units}^2$$

AREA BETWEEN CURVE AND THE Y-AXIS

It is sometimes necessary to find the area between the function and the y- axis.

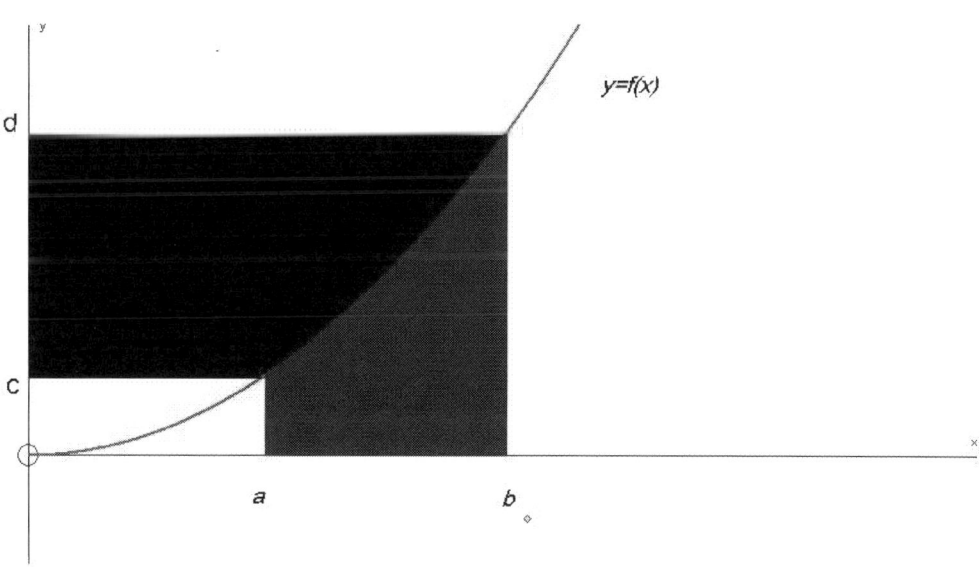

This is given as

$$\int_c^d f(y)dy .$$
or
$$\int_a^b (\text{right hand curve - left hand curve})dy .$$

It is not always possible to express the function y=f(x) in terms of x=f(y).

It may also be easier to calculate

$\int_a^b f(x)dx$ and subtract this as a composite area.

FUNDAMENTAL THEOREM OF CALCULUS

If $F(x)$ is the anti-derivative of $f(x)$
$$\int_a^b f(x)dx = F(b) - F(a) \quad (a \leq x \leq b)$$

Examples

Evaluate:

$\int_1^4 3x^2 dx$
$= \left[x^3\right]_1^4$
$= (4^3) - (1^3)$
$= 64 - 1$
$= 63 \text{ units}^2$

$\int_8^{15} \frac{5}{x^2} dx = \int_8^{15} 5x^{-2} dx$
$= \left[-5x^{-1}\right]_8^{15}$
$= \left[-\frac{5}{x}\right]_8^{15}$
$= \left(-\frac{5}{15}\right) - \left(-\frac{5}{8}\right)$
$= -\frac{1}{3} + \frac{5}{8}$
$= \frac{-8 + 15}{24}$
$= \frac{7}{24} \text{ units}^2$

Find the positive value of z :-

$\int_2^z (6x - 5)dx = 10$

$\Rightarrow \left[3x^2 - 5x\right]_2^z = 10$

$\Rightarrow (3z^2 - 5z) - (3 \times 2^2 - 5 \times 2) = 10$

$\Rightarrow (3z^2 - 5z) - (12 - 10) = 10$

$\Rightarrow 3z^2 - 5z - 12 = 0$

$\Rightarrow (3z + 4)(z - 3) = 0$

$\Rightarrow z = -\frac{4}{3} \text{ or } z = 3$

z = 3 is the positive solution

AREAS ENCLOSED BY THE GRAPH AND THE X – AXIS.

When calculating the area enclosed by a graph and the x-axis:-

Always draw a sketch,
calculate areas above and below x-axis separately,
and ignore negative signs and add.

Example

Calculate the area enclosed by the graph of y = x+2 and the x-axis for $-6 \leq x \leq 1$

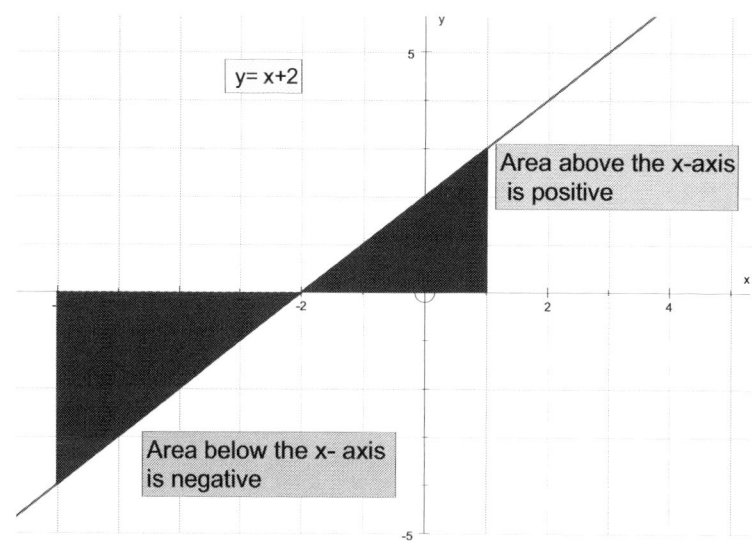

The graph cuts the x-axis at (-2 ,0)

Area below the x-axis

$$\int_{-6}^{-2}(x+2)dx = \left[\frac{x^2}{2} + 2x\right]_{-6}^{-2}$$

$$= \left(\frac{(-2)^2}{2} + 2\times(-2)\right) - \left(\frac{(-6)^2}{2} + 2\times(-6)\right)$$

$$= \left(\frac{4}{2} - 4\right) - \left(\frac{36}{2} - 12\right)$$

$$= -2 - 6$$

$$= -8 \text{ units}^2$$

Area above the x-axis

$$\int_{-2}^{1}(x+2)dx = \left[\frac{x^2}{2} + 2x\right]_{-2}^{1}$$

$$= \left(\frac{1^2}{2} + 2\times 1\right) - \left(\frac{(-2)^2}{2} + 2\times(-2)\right)$$

$$= \left(\frac{1}{2} + 2\right) - \left(\frac{4}{2} - 4\right)$$

$$= 2\frac{1}{2} + 2$$

$$= 4\frac{1}{2} \text{ units}^2$$

$$\text{Total area} = 8 + 4\frac{1}{2} = 12\frac{1}{2} \text{ units}^2$$

AREA BETWEEN TWO GRAPHS

The area between two graphs can be found by subtracting the area between the lower graph and the x-axis from the area between the upper graph and the x-axis.

Example
Calculate the area shaded between the graphs y= x+2 and y = x^2.

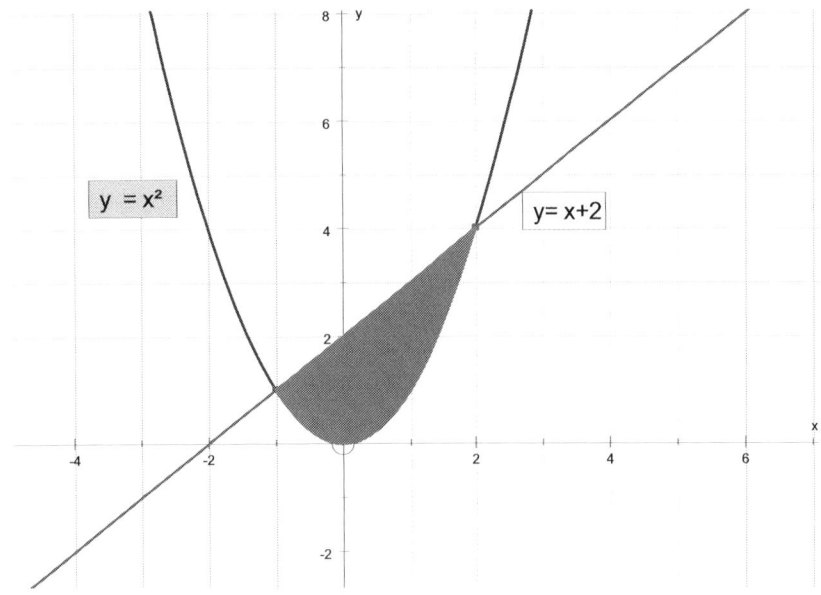

The graphs intersect at (-1 ,1) and (2,4).

Area between upper curve and x-

$$\int_{-1}^{2}(x+2)dx = \left[\frac{x^2}{2} + 2x\right]_{-1}^{2}$$

$$= \left(\frac{2^2}{2} + 2\times 2\right) - \left(\frac{(-1)^2}{2} + 2\times(-1)\right)$$

$$= \left(\frac{4}{2} + 4\right) - \left(\frac{1}{2} - 2\right)$$

$$= 6 + 1\frac{1}{2}$$

$$= 7\frac{1}{2} \text{ units}^2$$

Area between lower curve and x- axis

$$\int_{-1}^{2} x^2 dx = \left[\frac{x^3}{3}\right]_{-1}^{2}$$

$$= \left(\frac{2^3}{3}\right) - \left(\frac{(-1)^3}{3}\right)$$

$$= \left(\frac{8}{3}\right) - \left(\frac{-1}{3}\right)$$

$$= \frac{9}{3}$$

$$= 3 \text{ units}^2$$

Shaded area $= 7\frac{1}{2} - 3 = 4\frac{1}{2}$ units^2

FORMULA FOR AREA BETWEEN TWO GRAPHS

The area enclosed between the curves
$y = f(x)$ and $y = g(x)$
from $x = a$ to $x = b$ is

$$\int_a^b f(x)dx - \int_a^b g(x)dx = \int_a^b (f(x) - g(x))dx$$

$f(x) \geq g(x), \qquad a \leq x \leq b$

Area between curves is found by
$$\int_a^b \text{Upper curve} - \int_a^b \text{lower curve}$$

Example

Calculate the shaded area enclosed between the parabolas with equations

y = 1 +10x − 2x² and y = 1 +5x − x².

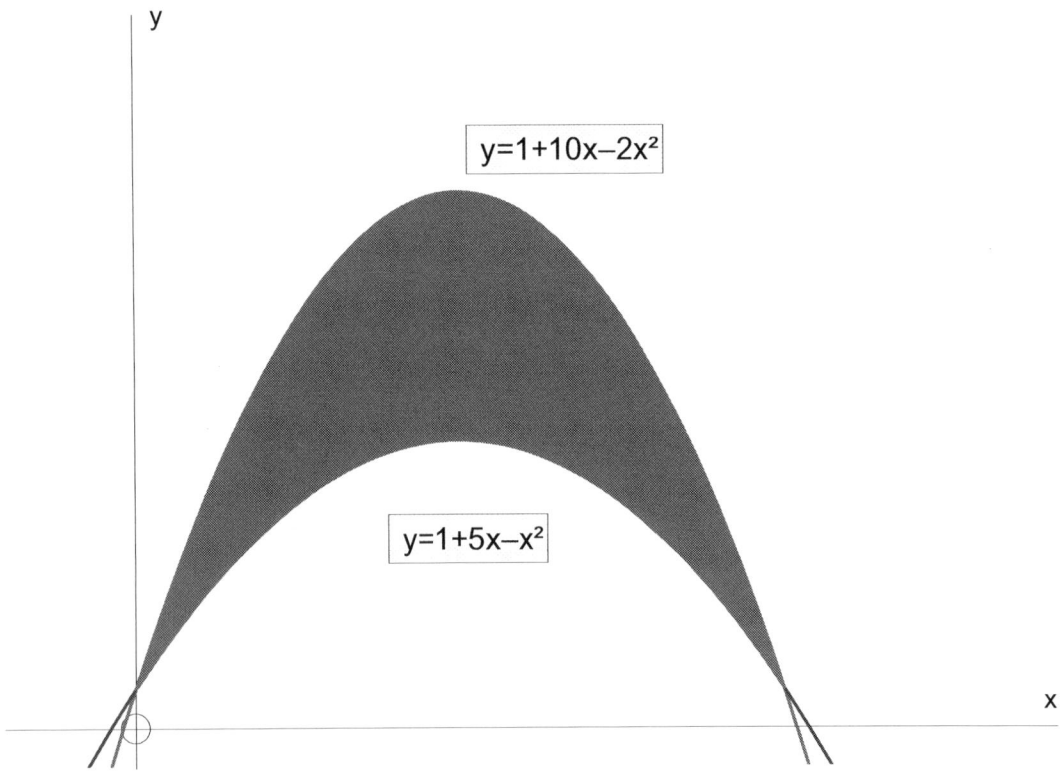

let $f(x) = 1 + 10x - 2x^2$
let $g(x) = 1 + 5x - x^2$
Curves intersect when $f(x) = g(x)$
$1 + 10x - 2x^2 = 1 + 5x - x^2$
$\Rightarrow 0 = x^2 - 5x$
$\Rightarrow 0 = x(x-5)$
$\Rightarrow x = 0$ and $x = 5$

Area between curves is found by

$$\int_a^b f(x)dx - \int_a^b g(x)dx = \int_a^b (f(x) - g(x))dx$$

so

$$\begin{aligned}
Area &= \int_0^5 ((1+10x-2x^2)-(1+5x-x^2))dx \\
&= \int_0^5 (5x-x^2)dx \\
&= \left[\frac{5x^2}{2} - \frac{x^3}{3}\right]_0^5 \\
&= \left(\frac{5\times 5^2}{2} - \frac{5^3}{3}\right) - 0 \\
&= \frac{125}{2} - \frac{125}{3} \\
&= \frac{125\times 3 - 125\times 2}{6} \\
&= \frac{125}{6} \\
&= 20\frac{5}{6} \text{ units}^2
\end{aligned}$$

INTEGRATING INVERSE TRIG FUNCTIONS.

$$\frac{d}{dx}\sin^{-1}(x) = \frac{1}{\sqrt{1-x^2}}$$

$$\Rightarrow \int \frac{dx}{\sqrt{1-x^2}} = \sin^{-1}(x) + C$$

$$\frac{d}{dx}\tan^{-1}(x) = \frac{1}{1+x^2}$$

$$\Rightarrow \int \frac{dx}{1+x^2} = \tan^{-1}(x) + C$$

$$\int \frac{dx}{\sqrt{a^2-x^2}} = \sin^{-1}\left(\frac{x}{a}\right) + C$$

$$\int \frac{dx}{a^2+x^2} = \frac{1}{a}\tan^{-1}\left(\frac{x}{a}\right) + C$$

Examples

Find the integral $\int \dfrac{dx}{\sqrt{4-x^2}}$

$$\int \frac{dx}{\sqrt{4-x^2}} = \int \frac{dx}{\sqrt{2^2-x^2}}$$

$$= \sin^{-1}\left(\frac{x}{2}\right) + C$$

Find the integral $\int \dfrac{3}{45+5x^2} dx$

$$\int \frac{3}{45+5x^2}dx = \int \frac{3dx}{5(9+x^2)}$$

$$= \frac{3}{5}\int \frac{dx}{(3^2+x^2)}$$

$$= \frac{3}{5}\cdot\frac{1}{3}\tan^{-1}\left(\frac{x}{3}\right) + C = \frac{1}{5}\tan^{-1}\left(\frac{x}{3}\right) + C$$

Find the integral $\int \dfrac{dx}{45+5x^2}$

$$\int \frac{dx}{45+5x^2} = \int \frac{dx}{5(9+x^2)}$$

$$= \frac{1}{5}\int \frac{dx}{(3^2+x^2)}$$

$$= \frac{1}{5}\cdot\frac{1}{3}\tan^{-1}\left(\frac{x}{3}\right) + C$$

$$= \frac{1}{15}\tan^{-1}\left(\frac{x}{3}\right) + C$$

INTEGRATION BY PARTS

The purpose here is to convert a tough integral into an easier one.
The following rule is used:

$$\int u \frac{dv}{dx} dx = uv - \int v \frac{du}{dx} dx$$

The rule comes from reversing the product rule for differentiation.

Steps:

Identify u and dv/dx
1. Find du/dx and v
2. Plug into formula.
3. Go again if necessary!

Tip : Pick u as the function which is easier to differentiate

Examples
Find the integral $\int x \sin x \, dx$

$\int x \sin x \, dx$ 　　　　　　　 let $u = x$ and dv/dx = sinx

　　　　　　　　　　　then $du/dx = 1$ and $v = \int \sin x \, dx = -\cos x$

$\int u \frac{dv}{dx} dx = uv - \int v \frac{du}{dx} dx$

$\int x \sin x \, dx = x(-\cos x) - \int (-\cos x) dx$

$\int x \sin x \, dx = -x \cos x + \int \cos x \, dx$

$\int x \sin x \, dx = -x \cos x + \sin x + C$

Find the integral $\int x^2 e^x dx$

$\int x^2 e^x dx$ let $u = x^2$ and $dv/dx = e^x$

         then $du/dx = 2x$ and $v = e^x$

$\int u \dfrac{dv}{dx} dx = uv - \int v \dfrac{du}{dx} dx$

$\int x^2 e^x dx = x^2 e^x - \int e^x \cdot 2x\, dx$

$\int x^2 e^x dx = x^2 e^x - \int 2x e^x dx$

Go again

$\int 2x e^x dx$ let $u = 2x$ and $dv/dx = e^x$

      then $du/dx = 2$ and $v = e^x$

$\int 2x e^x dx = 2x e^x - \int 2 e^x dx$

$\int x^2 e^x dx = x^2 e^x - (2x e^x - \int 2 e^x dx)$

$\int x^2 e^x dx = x^2 e^x - 2x e^x + 2\int e^x dx$

$\int x^2 e^x dx = x^2 e^x - 2x e^x + 2 e^x + C$

Find the integral $\int x \sqrt[3]{2-x}\, dx$

$\int x \sqrt[3]{2-x}\, dx$ let $u = x$ and $dv/dx = (2-x)^{1/3}$

         then $du/dx = 1$ and $v = -\dfrac{3(2-x)^{4/3}}{4}$

$\int u \dfrac{dv}{dx} dx = uv - \int v \dfrac{du}{dx} dx$

$\int x \sqrt[3]{2-x}\, dx = \dfrac{-3x(2-x)^{4/3}}{4} + \int \dfrac{3(2-x)^{4/3}}{4} dx$

$\int x \sqrt[3]{2-x}\, dx = \dfrac{-3x(2-x)^{4/3}}{4} + \dfrac{3}{4} \int (2-x)^{4/3} dx$

$\int x \sqrt[3]{2-x}\, dx = \dfrac{-3x(2-x)^{4/3}}{4} + \dfrac{3}{4} \cdot \dfrac{-3}{7}(2-x)^{7/3}$

$\int x \sqrt[3]{2-x}\, dx = \dfrac{-3x(2-x)^{4/3}}{4} - \dfrac{9(2-x)^{7/3}}{28}$

INTEGRATING RATIONAL FUNCTIONS

Examples

Find the integral $\int \dfrac{4x+12}{x(x+4)} dx$

$$\dfrac{4x+12}{x(x+4)} = \dfrac{A}{x} + \dfrac{B}{x+4}$$
$$= \dfrac{A(x+4)+Bx}{x(x+4)}$$

comparing coefficients

$4x+12 \equiv A(x+4)+Bx$

Let $x = -4$ \qquad Let $x = 0$

$\Rightarrow -4 = -4B$ \qquad $\Rightarrow 12 = 4A$

$\Rightarrow B = 1$ \qquad $\Rightarrow A = 3$

$\dfrac{4x+12}{x(x+4)} = \dfrac{3}{x} + \dfrac{1}{(x+4)}$

$\int \dfrac{4x+12}{x(x+4)} dx = \int \dfrac{3}{x} + \dfrac{1}{(x+4)} dx$

$= 3\int \dfrac{1}{x} dx + \int \dfrac{1}{(x+4)} dx$

$= 3\ln|x| + \ln|x+4| + C$

$= \ln|x^3(x+4)| + C$

(letting $C = \ln K$)

$= \ln|kx^3(x+4)|$

Find the integral $\int \dfrac{xdx}{(x+1)^2}$

$$\dfrac{x}{(x+1)^2} = \dfrac{A}{(x+1)} + \dfrac{B}{(x+1)^2}$$
$$= \dfrac{A(x+1)+B}{(x+1)^2}$$

comparing coefficients

$x \equiv A(x+1)+B$

Let $x = -1$ \qquad\qquad Let $x = 1$

$\Rightarrow -1 = B$ \qquad\qquad $\Rightarrow 1 = 2A + B$

\qquad\qquad\qquad\qquad $\Rightarrow 2A = 2$

\qquad\qquad\qquad\qquad $\Rightarrow A = 1$

$$\dfrac{x}{(x+1)^2} = \dfrac{1}{(x+1)} - \dfrac{1}{(x+1)^2}$$

$$\int \dfrac{xdx}{(x+1)^2} = \int \dfrac{1}{(x+1)} - \dfrac{1}{(x+1)^2}\,dx$$
$$= \int \dfrac{1}{(x+1)}dx - \int \dfrac{1}{(x+1)^2}dx$$

$$= \ln|x+1| + \dfrac{1}{(x+1)} + C$$

Find the integral $\int \dfrac{(2x-1)dx}{(x-1)(x-2)(x+1)}$

$$\dfrac{(2x-1)}{(x-1)(x-2)(x+1)} = \dfrac{A}{(x-1)} + \dfrac{B}{(x-2)} + \dfrac{C}{(x+1)}$$
$$= \dfrac{A(x-2)(x+1) + B(x-1)(x+1) + C(x-1)(x-2)}{(x-1)(x-2)(x+1)}$$

comparing coefficients
$$2x - 1 \equiv A(x-2)(x+1) + B(x-1)(x+1) + C(x-1)(x-2)$$

let $x = 2$	let $x = 1$	let $x = -1$
$3 = 3B$	$1 = -2A$	$-3 = 6C$
$\Rightarrow B = 1$	$\Rightarrow A = -1/2$	$\Rightarrow C = -1/2$

$$\dfrac{(2x-1)}{(x-1)(x-2)(x+1)} = \dfrac{-1}{2(x-1)} + \dfrac{1}{(x-2)} - \dfrac{1}{2(x+1)}$$

$$\int \dfrac{(2x-1)dx}{(x-1)(x-2)(x+1)} = \int \dfrac{-1}{2(x-1)} + \dfrac{1}{(x-2)} - \dfrac{1}{2(x+1)} dx$$

$$= \int \dfrac{-1}{2(x-1)} dx + \int \dfrac{1}{(x-2)} dx - \int \dfrac{1}{2(x+1)} dx$$

$$= -\dfrac{1}{2}\int \dfrac{dx}{(x-1)} + \int \dfrac{dx}{(x-2)} - \dfrac{1}{2}\int \dfrac{dx}{(x+1)}$$

$$= -\dfrac{1}{2}\ln|x-1| + \ln|x-2| - \dfrac{1}{2}\ln|x+1| + C$$

$$= \ln\left|\dfrac{(x-2)}{(x-1)^{1/2}(x+1)^{1/2}}\right| + C$$

$$= \ln\left|\dfrac{(x-2)}{(x^2-1)^{1/2}}\right| + C$$

(*letting* $C = \ln K$)

$$= \ln\left|\frac{k(x-2)}{(x^2-1)^{1/2}}\right|$$

Find the integral $\int \dfrac{x^3+x^2-4x+1}{x^2+x-6}\,dx$

$$\begin{array}{r}x\\x^2+x-6\overline{\smash{)}x^3+x^2-4x+1}\\x^3+x^2-6x\\\hline 2x+1\end{array}$$

resolving the rational portion

$$\frac{2x+1}{x^2+x-6} = \frac{A}{(x-2)} + \frac{B}{(x+3)}$$

$$\frac{x^3+x^2-4x+1}{x^2+x-6} = X + \frac{2x+1}{x^2+x-6}$$

$$= \frac{A(x+3)+B(x-2)}{(x-2)(x+3)}$$

comparing coefficients

$2x+1 \equiv A(x+3)+B(x-2)$

let $x = -3$ let $x = 2$

$-5 = -5B$ $5 = 5A$

$\Rightarrow B = 1$ $\Rightarrow A = 1$

$$\frac{x^3+x^2-4x+1}{x^2+x-6} = X + \frac{1}{(x-2)} + \frac{1}{(x+3)}$$

DIFFERENTIAL EQUATIONS

MAC 1.4 Applying calculus skills to solving differential equations.

Apps 1.5 Applying algebraic and calculus skills to problems.

These equations, containing a derivative, involve rates of change – so often appear in an engineering or scientific context. Solving the equation involves integration.

The order of a differential equation is given by the highest derivative used.

The degree of a differential equation is given by the degree of the power of the highest derivative used.

Examples :-

$\dfrac{dy}{dx} = 5x+6$ has order 1 and is 1st degree

$\dfrac{d^2y}{dx^2} - \dfrac{dy}{dx} + 9 = 0$ has order 2 and is 1st degree

$\dfrac{d^3y}{dx^3} = 78xe^8$ has order 3 and is 1st degree

$\left(\dfrac{dy}{dx}\right)^2 = 5x+6$ has order 1 and is 2nd degree

$\left(\dfrac{d^2y}{dx^2}\right)^7 - \left(\dfrac{dy}{dx}\right)^{20} + 9 = 0$ has order 2 and is 7th degree

$\left(\dfrac{d^3y}{dx^3}\right)^2 = 78xe^8$ has order 3 and is 2nd degree

FIRST ORDER DIFFERENTIAL EQUATIONS

Solving by direct integration.

The general solution of differential equations of the form $\dfrac{dy}{dx} = f(x)$ can be found using direct integration.

Substituting the values of the initial conditions will give particular solutions.

e.g.

$\dfrac{dy}{dx} = 6x$

$dy = 6x\,dx$

$\int dy = \int 6x\,dx$

$y = \dfrac{6x^2}{2} + c$

$y = 3x^2 + c$

The general solution of the differential equation

is $y = 3x^2 + c$

A particular solution of the differential equation

is $y = 3x^2 + 6$

Examples

Find the particular solution of the differential equation $\dfrac{dy}{dx} = 4x - 2$ given y = 5 when x = 3.

$\dfrac{dy}{dx} = 4x - 2$

$\Rightarrow dy = (4x - 2)dx$

$\Rightarrow \int dy = \int 4x - 2 dx$

$\Rightarrow y = 2x^2 - 2x + c$

$y = 2x^2 - 2x + c$ is the general solution

When $y = 5$, $x = 3$

$5 = 2 \times 3^2 - 2 \times 3 + c$

$5 = 18 - 6 + c$

$5 - 12 = c$

$c = -7$

The particular solution for the given initial conditions is $y = 2x^2 - 2x - 7$

A straight line with gradient 2 passes through the point (1,3).

Find the equation of the line.

$\dfrac{dy}{dx} = 2$

$\Rightarrow dy = 2dx$

$\Rightarrow \int dy = \int 2dx$

$\Rightarrow y = 2x + c$

When $x = 1$, $y = 3$

$3 = 2 + c$

$c = 1$

The equation is $y = 2x + 1$

VARIABLES SEPERABLE

A variables separable differential equation is one in which the equation can be written with all the terms for one variable on one side of the equation, and the other terms on the other side.

$$\frac{dy}{dx} = f(y)$$
$$\Rightarrow \frac{dy}{f(y)} = dx$$
$$\Rightarrow \int \frac{dy}{f(y)} = \int dx$$

$$\frac{dy}{dx} = f(x)g(y)$$
$$\Rightarrow \frac{dy}{g(y)} = f(x)dx$$
$$\Rightarrow \int \frac{dy}{g(y)} = \int f(x)dx$$

Examples

Find the general solution of the differential equation $y\frac{dy}{dx} = x$

$y\frac{dy}{dx} = x$

$\Rightarrow ydy = xdx$

$\Rightarrow \int ydy = \int xdx$

$\Rightarrow \frac{y^2}{2} = \frac{x^2}{2} + c$ let $C = 2c$

$\Rightarrow y^2 = x^2 + C$

Find the general solution of the differential equation $x\dfrac{dy}{dx} = y$

$x\dfrac{dy}{dx} = y$

$\Rightarrow xdy = ydx$

$\Rightarrow \dfrac{dy}{y} = \dfrac{dx}{x}$

$\Rightarrow \int \dfrac{dy}{y} = \int \dfrac{dx}{x}$

$\Rightarrow \ln|y| = \ln|x| + c$

$\Rightarrow \ln|y| = \ln|x| + \ln k \quad$,where $c = \ln k$

$\Rightarrow y = kx$

Find the particular solution of the differential equation $x\dfrac{dy}{dx} = y(y+1)$ given y = 2 when x = 1.

$x\dfrac{dy}{dx} = y(y+1)$

$\Rightarrow xdy = y(y+1)dx$

$\Rightarrow \dfrac{dy}{y(y+1)} = \dfrac{dx}{x}$

$\Rightarrow \int \dfrac{dy}{y(y+1)} = \int \dfrac{dx}{x}$

Partial fractions are required to break the left hand side of the equation into a form which can be integrated.

$1 \equiv A(y+1) + By$

let $y = 0$ $\qquad$ let $y = -1$

$A = 1$ $\qquad\qquad$ $B = -1$

$\Rightarrow \int \dfrac{dy}{y(y+1)} = \int \dfrac{1}{y}dy - \int \dfrac{1}{y+1}dy$

$$\int \frac{dy}{y(y+1)} = \int \frac{dx}{x}$$

$$\Rightarrow \int \frac{1}{y}dy - \int \frac{1}{y+1}dy = \int \frac{dx}{x}$$

$$\Rightarrow \ln|y| - \ln|y+1| = \ln|x| + c \qquad \text{let } c = \ln k$$

$$\Rightarrow \ln|y| - \ln|y+1| = \ln|x| + \ln k$$

$$\Rightarrow \ln\left|\frac{y}{y+1}\right| = \ln|kx|$$

$$\Rightarrow \left|\frac{y}{y+1}\right| = |kx|$$

when $y = 2, x = 1$

$$\frac{2}{2+1} = k$$

$$\Rightarrow \frac{2}{3} = k$$

$$\Rightarrow \left|\frac{y}{y+1}\right| = \left|\frac{2}{3}x\right|$$

APPLICATIONS OF DIFFERENTIAL EQUATIONS

Differential equations are dynamic, involving instantaneous rates of change.

Examples :-

ROCKET SCIENCE

A toy rocket, consisting of casing and fuel, is launched at time t=0.
The initial total mass of the rocket is m_o.
The fuel burns such that the mass of the rocket satisfies the equation

$$\frac{dm}{dt} = -\frac{1}{4}t$$

Given that the mass of the casing $=1/2\ m_o$, for how long does the rocket's fuel burn?

$$\frac{dm}{dt} = -\frac{1}{4}t$$

$$\Rightarrow dm = -\frac{1}{4}t\,dt$$

$$\Rightarrow \int dm = \int -\frac{1}{4}t\,dt$$

$$\Rightarrow m = -\frac{t^2}{8} + c$$

When $t = 0$, $m = m_o$

$m_o = c$

$$\Rightarrow m = -\frac{t^2}{8} + m_o$$

$$\Rightarrow m = m_o - \frac{t^2}{8}$$

The rocket has spent its fuel when $m = \frac{1}{2}m_o$

$$\Rightarrow m(T) = \frac{1}{2}m_o$$

but $m(T) = m_o - \frac{t^2}{8}$

$$\Rightarrow \frac{1}{2}m_o = m_o - \frac{t^2}{8}$$

$$\Rightarrow \frac{t^2}{8} = \frac{1}{2}m_o$$

$$\Rightarrow t^2 = 4m_o$$

$$\Rightarrow t = \sqrt{4m_o}$$

$$\Rightarrow t = 2\sqrt{m_o}$$

NEWTON'S LAW OF COOLING

The rate at which an object cools is proportional to the difference between its temperature and that of its surroundings.

The temperature θ°C of a cup of coffee t minutes after it was made is given by the equation

$$\frac{d\theta}{dt} = -k(\theta - 20)$$

The initial temperature of the drink is 70°C.
5 minutes later, its temperature is 50°C.
When is the temperature 40°C?

$\dfrac{d\theta}{dt} = -k(\theta - 20)$

$\Rightarrow \dfrac{d\theta}{(\theta - 20)} = -k\,dt$

$\Rightarrow \displaystyle\int \dfrac{d\theta}{(\theta - 20)} = \int -k\,dt$

$\Rightarrow \ln|\theta - 20| = -kt + c$

$\Rightarrow \ln\left|\dfrac{\theta - 20}{C}\right| = -kt$

$\Rightarrow \theta - 20 = Ce^{-kt}$

When $t = 0$, $\theta = 70$

$\Rightarrow 50 = C$

$\Rightarrow \theta = 50e^{-kt} + 20$

When $t = 5$, $\theta = 50$

$\Rightarrow 50 = 50e^{-5k} + 20$

$\Rightarrow 30 = 50e^{-5k}$

$\Rightarrow \dfrac{\ln(3/5)}{-5} = k$

$\Rightarrow \theta = 50e^{-0.1021t} + 20$

$40 = 50e^{-0.1021t} + 20$

$\Rightarrow 20 = 50e^{-0.1021t}$

$\Rightarrow \dfrac{\ln(2/5)}{-0.1021} = t$

$\Rightarrow t = 8.97$ minutes after it was made

GROWTH & DECAY

The rate of growth of a population P is given by $2P$ per day.
Initially, the population is 100.
Find how long it takes for the population to double.

$$\frac{dP}{dt} = 2P$$

$$\Rightarrow \frac{dP}{2P} = dt$$

$$\Rightarrow \int \frac{dP}{2P} = \int dt$$

$$\Rightarrow \frac{1}{2}\ln|P| = t + c$$

$$\Rightarrow \ln\left|\frac{P^{\frac{1}{2}}}{C}\right| = t$$

$$\Rightarrow P^{\frac{1}{2}} = Ce^t$$

When $t = 0$, $P = 100$

$$\Rightarrow = C$$

$$\Rightarrow P^{\frac{1}{2}} = \sqrt{100}e^t$$

$$\Rightarrow P = 100e^{2t}$$

When $P = 200$

$$200 = 100e^{2t}$$

$$\Rightarrow 2 = e^{2t}$$

$$\Rightarrow \frac{\ln 2}{2} = t$$

$$\Rightarrow t = 0.346 \text{ days}$$

$$\Rightarrow t = 0.346 \times 24 = 8.3 \text{ hours}$$

KINEMATICS

The acceleration a ms^{-2} of a body moving in a straight line is given as $a = 3x^2$, where x is the distance from the origin in metres.

When x = 1, the body has velocity 2 ms^{-1}

Find the velocity when x = 2

$a = v\dfrac{dv}{dx} = 3x^2$

$\Rightarrow vdv = 3x^2 dx$

$\Rightarrow \int vdv = \int 3x^2 dx$

$\Rightarrow \dfrac{1}{2}v^2 = x^3 + c$

When $v = 2$, $x = 1$

$\Rightarrow 2 = 1 + c$

$\Rightarrow c = 1$

$\Rightarrow \dfrac{1}{2}v^2 = x^3 + 1$

When $x = 2$

$\Rightarrow \dfrac{1}{2}v^2 = 2^3 + 1$

$\Rightarrow v = \sqrt{18}$

$\Rightarrow v = 3\sqrt{2}$ ms-1

VOLUMES OF SOLID OF REVOLUTION

If a curve is rotated about an axis which it does not cut, the area between the curve and the axis will generate a *solid of revolution*.

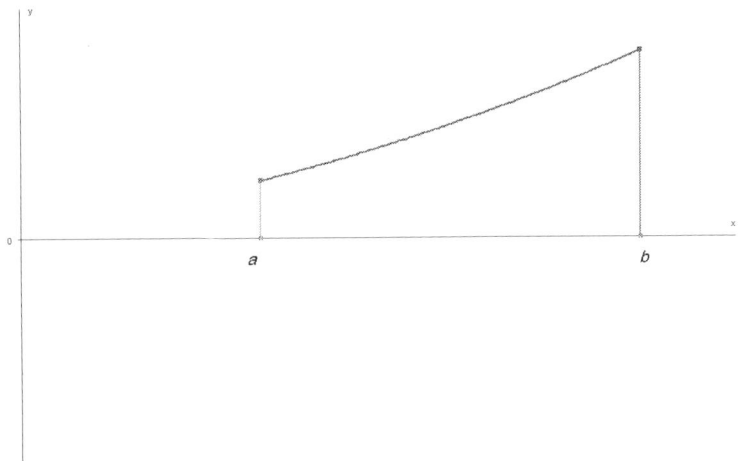

The volume generated is the limit of the sums of the volumes of cylinders formed by rotating rectangles.

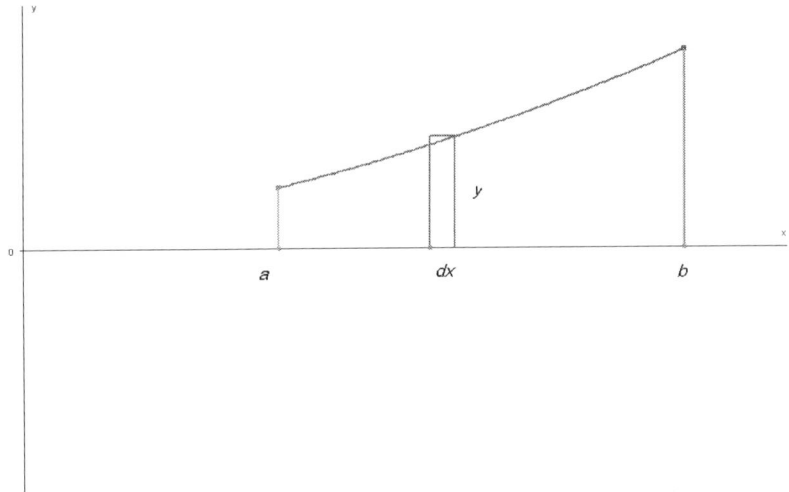

If rectangle dx.y is rotated, a cylinder is formed with height dy and radius y.

The volume of this cylinder is $V = \pi r^2 h = \pi y^2 dx$

The volume of the solid of revolution is

$$\lim_{\delta x \to 0} \sum_{x=a}^{b} \pi y^2 dx = \int_a^b \pi y^2 dx$$

Example

Find the volume and name the shape generated by the rotation through 360° of the positive branch of the circle with equation $x^2+y^2=a^2$ about the x-axis.

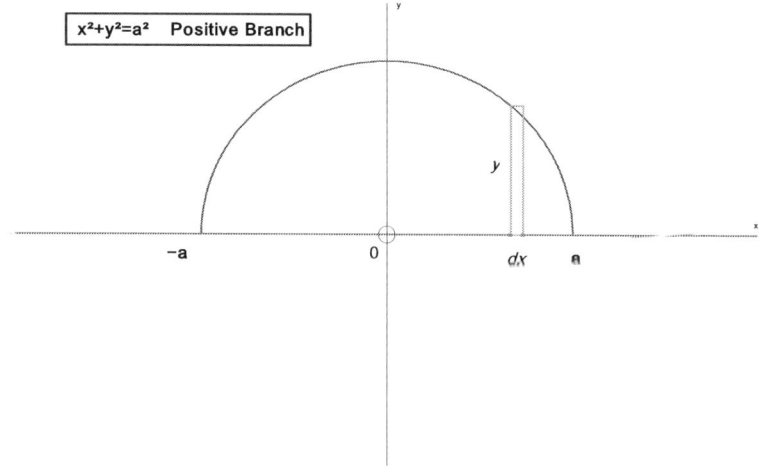

$$V = \lim_{\delta x \to 0} \sum_{x=-a}^{a} \pi y^2 dx = \int_{-a}^{a} \pi y^2 dx \qquad x^2 + y^2 = a^2$$

$$= \pi \int_{-a}^{a} y^2 dx \qquad y^2 = a^2 - x^2$$

$$= \pi \int_{-a}^{a} (a^2 - x^2) dx$$

$$= \pi \left[a^2 x - \frac{x^3}{3} \right]_{-a}^{a}$$

$$= \pi \{ (a^2 a - \frac{a^3}{3}) - (a^2(-a) - \frac{(-a)^3}{3}) \}$$

$$= \pi \{ (a^3 - \frac{a^3}{3}) - (-a^3 + \frac{a^3}{3}) \}$$

$$= \pi \{ a^3 - \frac{a^3}{3} + a^3 - \frac{a^3}{3} \}$$

$$= 2\pi (a^3 - \frac{a^3}{3})$$

$$= \frac{4\pi a^3}{3}$$

This is a sphere - the volume is of the form $\frac{4}{3}\pi r^3$

Example

The portion of the curve $y=2x^2$ between $y=1$ and $y=3$ is rotated through $360°$ about the y-axis.

Find the volume of the resulting solid.

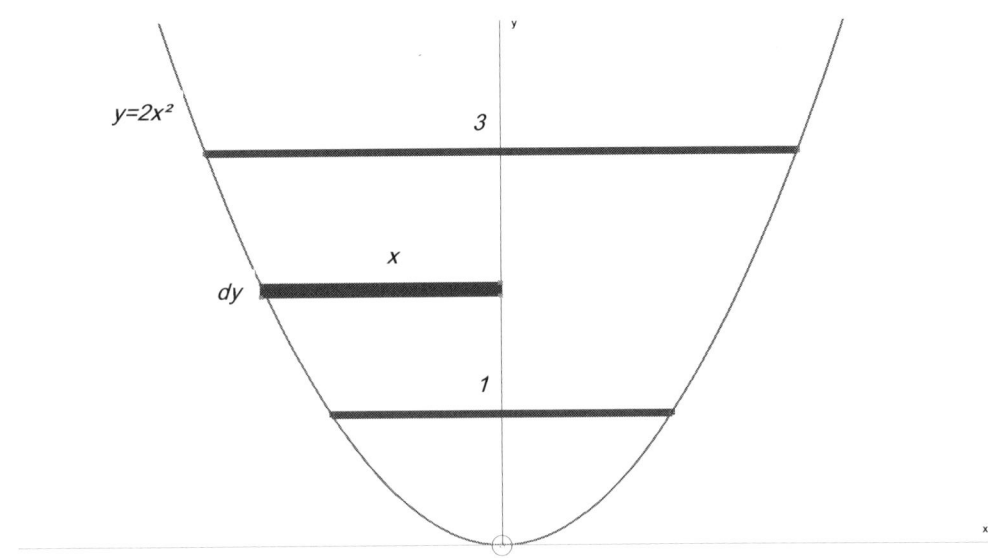

$$V = \lim_{\delta x \to 0} \sum_{y=1}^{3} \pi x^2 dy = \int_{1}^{3} \pi x^2 dy \qquad y = 2x^2$$

$$= \int_{1}^{3} \pi \frac{y}{2} dy \qquad x^2 = \frac{y}{2}$$

$$= \frac{\pi}{2} \int_{1}^{3} y \, dy$$

$$= \frac{\pi}{2} \left[\frac{y^2}{2} \right]_{1}^{3}$$

$$= \frac{\pi}{2} \{ \frac{3^2}{2} - \frac{1^2}{2} \}$$

$$= \frac{\pi}{2} \{ \frac{9}{2} - \frac{1}{2} \}$$

$$= 2\pi$$

APPLICATIONS OF INTEGRATION

Example

A car starts from rest and its acceleration t seconds after the start is $1/10(20-t)$ ms^{-2}. What is its speed after 20 seconds?

Remember
$$a = \frac{dv}{dt}, \quad v = \int a\,dt$$
$$v = \frac{ds}{dt}, \quad s = \int v\,dt$$

$$a = \frac{dv}{dt} = \frac{1}{10}(20-t)$$

$$v = \int a\,dt$$

$$= \int \frac{1}{10}(20-t)\,dt$$

$$= \frac{1}{10}\int(20-t)\,dt$$

$$= \frac{1}{10}(20t - \frac{t^2}{2}) + C$$

$$= 2t - \frac{t^2}{20} + C$$

When $t=0$, $v=0$ so $C = 0$

so
$$v = 2t - \frac{t^2}{20}$$

At $t = 20$

$$v = 2 \times 20 - \frac{20^2}{20}$$
$$= 40 - 20$$
$$= 20 \text{ ms}^{-1}$$

The car ceases to accelerate and continues at this uniform velocity. What is the total distance covered 30 seconds after it started?

$$s = \int v\,dt$$

$$= \int 2t - \frac{t^2}{20}\,dt$$

$$= \left[(t^2 - \frac{t^3}{60})\right]_0^{20}$$

$$= \left((20 \times 20 - \frac{20^3}{60}) - 0\right)$$

$$= 400 - \frac{8000}{60}$$

$$= 266\frac{2}{3} \text{ m}$$

total distance $= 266\frac{2}{3}$ m for the first 20 seconds
+ 200 m for the next 10 seconds at 20 m/s
$= 466\frac{2}{3}$ m

297

LINEAR DIFFERENTIAL EQUATIONS

These are first degree differential equations.

$$a_n(x)\frac{d^n y}{dx^n} + a_{n-1}(x)\frac{d^{n-1} y}{dx^{n-1}} + \ldots + a_1(x)\frac{dy}{dx} + a_0(x)y = f(x)$$

describes a general linear differential equation of order n, where $a_n(x)$, $a_{n-1}(x)$, etc and $f(x)$ are given functions of x or constants.

Louis Arbogast introduced the differential operator
$D = d/dx$, which simplifies the general equation to

$$a_n(x)D^n y + a_{n-1}(x)a_n(x)D^{n-1}y + \ldots + a_1(x)Dy + a_0(x)y = f(x)$$

or

$$\sum_{i=0}^{n} a_i(x)D^i y = f(x)$$

If $f(x) = 0$, the equation is called homogeneous.
If $f(x) \neq 0$, the equation is non-homogeneous.

FIRST ORDER LINEAR DIFFERENTIAL EQUATIONS

To solve equations of the form

$$a(x)\frac{dy}{dx} + b(x)y = g(x)$$

express in standard form

$$\boxed{\frac{dy}{dx} + P(x)y = Q(x)}$$

where P and Q are functions of x or constants

Multiply both sides by the Integrating Factor

$$\boxed{e^{\int P(x)dx}}$$

Write $\dfrac{d}{dx}(ye^{\int P(x)dx}) = Q(x)e^{\int P(x)dx}$

Integrate the right hand side, use integration by parts if necessary.

$$\boxed{ye^{\int P(x)dx} = \int e^{\int P(x)dx} Q(x)dx + C}$$

Divide both sides by the integrating factor.
This gives the General solution
Use any initial conditions to find particular solutions.

Example

Find a general solution of the equation

$$x^2 \frac{dy}{dx} + 4xy = x^3 - 1$$

$$x^2 \frac{dy}{dx} + 4xy = x^3 - 1$$

$$\Rightarrow \frac{dy}{dx} + \frac{4xy}{x^2} = \frac{x^3 - 1}{x^2}$$

$$\Rightarrow \frac{dy}{dx} + \frac{4y}{x} = x - \frac{1}{x^2}$$

Compare to

$$\frac{dy}{dx} + P(x)y = Q(x)$$

$$P(x) = \frac{4}{x} \qquad\qquad \int P(x) = \int \frac{4}{x} = 4\ln|x| = \ln|x|^4$$

$$\therefore \frac{d}{dx}(yx^4) = x^4\left(x - \frac{1}{x^2}\right) \qquad \text{Integrating Factor is}$$

$$\qquad\qquad\qquad\qquad\qquad\qquad\qquad e^{\int P(x)} = e^{\ln|x|^4} = x^4$$

$$\Rightarrow yx^4 = \int x^4\left(x - \frac{1}{x^2}\right)dx$$

$$\Rightarrow yx^4 = \int x^5 - x^2 \, dx$$

$$\Rightarrow yx^4 = \frac{x^6}{6} - \frac{x^3}{3} + c$$

$$\Rightarrow y = \frac{x^6}{6x^4} - \frac{x^3}{3x^4} + \frac{c}{x^4}$$

$$\Rightarrow y = \frac{x^2}{6} - \frac{1}{3x} + \frac{c}{x^4} \quad \text{is a general solution}$$

Find a general solution of the equation

$$\frac{dy}{dx} + \frac{3y}{x+2} = x+2$$

where $x \neq 2$, and hence find the particular solution for $y = 1$ when $x = -1$

$$\frac{dy}{dx} + \frac{3y}{x+2} = x+2$$

$$P(x) = \frac{3}{x+2}$$

$$\int P(x) = \int \frac{3}{x+2} = 3\ln|x+2| = \ln|x+2|^3$$

Integrating Factor is

$$e^{\int P(x)} = e^{\ln|x+2|^3} = (x+2)^3$$

$$\therefore \frac{d}{dx}(y(x+2)^3) = (x+2)^3(x+2)$$

$$\Rightarrow y(x+2)^3 = \int (x+2)^4 dx$$

$$\Rightarrow y(x+2)^3 = \frac{(x+2)^5}{5} + c$$

$$\Rightarrow y = \frac{(x+2)^5}{5(x+2)^3} + c$$

$$\Rightarrow y = \frac{(x+2)^2}{5} + \frac{c}{(x+2)^3} \text{ is a general solution}$$

When $x = -1, y = 1$

$$1 = \frac{(-1+2)^2}{5} + \frac{c}{(-1+2)^3}$$

$$1 = \frac{1}{5} + c$$

$$\Rightarrow c = \frac{4}{5}$$

The particular solution is

$$y = \frac{(x+2)^2}{5} + \frac{4}{5(x+2)^3}$$

SECOND ORDER LINEAR DIFFERENTIAL EQUATIONS

To solve equations of the form

$$a(x)\frac{d^2y}{dx^2} + b(x)\frac{dy}{dx} + c(x)y = 0$$

Write down the auxiliary equation

$$\boxed{am^2 + bm + c = 0}$$

Examine the discriminant of the auxiliary equation.

For real and distinct roots,
m_1 and m_2, the general solution is

$$\boxed{y = Ae^{m_1x} + Be^{m_2x}}$$

For real and equal roots,
the general solution is

$$\boxed{y = Ae^{mx} + Bxe^{mx}}$$

For complex conjugate roots,
$m_1 = p + iq$ and $m_2 = p - iq$,
the general solution is

$$\boxed{y = e^{px}(C\cos qx + D\sin qx)}$$

Use any initial conditions to find the particular solution.

Example

Find the general solution of the equation

$$\frac{d^2y}{dx^2} + \frac{dy}{dx} - 12y = 0$$

and the particular solution for which
y = 7 when x=0 and dy/dx = 7

$\frac{d^2y}{dx^2} + \frac{dy}{dx} - 12y = 0$

has auxillary equation

$m^2 + m - 12 = 0$

whose discriminant is

$1 - (4 \times 1 \times (-12)) = 49$

∴ *real* and distinct roots exist

$m^2 + m - 12 = 0$
$\Rightarrow (m+4)(m-3) = 0$
$\Rightarrow m = -4$ or $m = 3$

The general solution is

$y = Ae^{-4x} + Be^{3x}$

$y = Ae^{-4x} + Be^{3x}$

$\therefore \frac{dy}{dx} = -4Ae^{-4x} + 3Be^{3x}$

When x=0, dy/dx = 7
$\Rightarrow 7 = -4Ae^0 + 3Be^0$
$\Rightarrow 7 = -4A + 3B$

When x=0, y= 7
$\Rightarrow 7 = Ae^0 + Be^0$
$\Rightarrow 7 = A + B$

Solving simultaneously

$7 = A + B$
$7 = -4A + 3B$
$\Rightarrow 7 = -4A + 3(7 - A)$
$\Rightarrow 7 = -7A + 21$

$A = 2$
$B = 5$

the particular solution is $y = 2e^{-4x} + 5e^{3x}$

Example

Find the general solution of the equation

$$\frac{d^2y}{dx^2} - 10\frac{dy}{dx} + 25y = 0$$

and the particular solution for y=0 and dy/dx = 3 when x=0

$$\frac{d^2y}{dx^2} - 10\frac{dy}{dx} + 25y = 0$$

has auxillary equation

$m^2 - 10m + 25 = 0$

whose discriminant is

$100 - (4 \times 1 \times (25)) = 0$

∴ *real* and equal roots exist

$m^2 - 10m + 25 = 0$
$\Rightarrow (m-5)(m-5) = 0$
$\Rightarrow m = 5$

The general solution is

$y = Ae^{5x} + Bxe^{5x}$

$y = Ae^{5x} + Bxe^{5x}$

$\therefore \frac{dy}{dx} = 5Ae^{5x} + Be^{5x} + 5Bxe^{5x}$

When x=0, dy/dx = 3

$\Rightarrow 3 = 5Ae^0 + Be^0$
$\Rightarrow 3 = 5A + B$

When x=0, y= 0

$\Rightarrow 0 = Ae^0$
$\Rightarrow 0 = A$

Solving simultaneously

$0 = A$
$3 = 5A + B$
$\Rightarrow 3 = B$

$y = Ae^{5x} + Bxe^{5x}$

Substituting for $A = 0$ and $B = 3$

the particular solution is $y = 3xe^{5x}$

Example

Find the general solution of the equation

$$\frac{d^2y}{dx^2} - 6\frac{dy}{dx} + 13y = 0$$

$$\frac{d^2y}{dx^2} - 6\frac{dy}{dx} + 13y = 0$$

has auxillary equation
$m^2 - 6m + 13 = 0$
whose discriminant is
$36 - (4 \times 1 \times 13) = -16$
∴ *complex* conjugate roots exist

$m^2 - 6m + 13 = 0$
$\Rightarrow m = \dfrac{6 \pm 4i}{2} = 3 \pm 2i$

The general solution is

$$y = e^{3x}(C\cos 2x + D\sin 2x)$$

SECOND ORDER NON HOMOGENEOUS DIFFERENTIAL EQUATIONS

The solution to equations of the form

$$a(x)\frac{d^2y}{dx^2} + b(x)\frac{dy}{dx} + c(x)y = Q(x)$$

has two parts, the complementary function (CF) and the particular integral (PI).

$$Q(x) = CF + PI$$

The CF is the general solution as described above for solving homogeneous equations.

The Particular Integral is found by substituting a form similar to Q(x) into the left-hand side equation, and equating co-efficients.

If Q(x) is a linear function, try $y = Cx + D$
If Q(x) is quadratic, try $Cx^2 + Dx + E$
If Q(x) is wave function, try $C\sin x + D\cos x$
If Q(x) is a constant, try $y = C$
If Q(x) is e^{kx}, try $y = C e^{kx}$

> The PI cannot have the same form as any of the terms in the CF, so care has to be taken to ensure that this is not the case.
>
> In such a situation, an extra x term is usually introduced to the PI.

A particular solution is found by substituting initial conditions into the general solution.

Do not just use the CF!!!

Example

Find the general solution of the equation

$$\frac{d^2y}{dx^2} - 3\frac{dy}{dx} + 2y = x+1$$

$$\frac{d^2y}{dx^2} - 3\frac{dy}{dx} + 2y = x+1$$

has auxillary equation
$m^2 - 3m + 2 = 0$
whose discriminant is
$9 - (4 \times 1 \times 2) = 1$
∴ *real* and distinct roots exist

$m^2 - 3m + 2 = 0$
$\Rightarrow m = 2$ or 1
The complementary function is
$y = Ae^{2x} + Be^x$

For the PI, try $y = Cx + D$

$$\therefore \frac{dy}{dx} = C \quad \text{and} \quad \frac{d^2y}{dx^2} = 0$$

$$\frac{d^2y}{dx^2} - 3\frac{dy}{dx} + 2y = x+1$$
$0 - 3C + 2(Cx + D) = x + 1$
$\Rightarrow 2Cx + 2D - 3C = x + 1$
Equating co-efficients,
$\quad 2C = 1 \quad$ and $\quad 2D-3C=1$
$\Rightarrow C=1/2 \quad$ and $\quad D =5/4$

The PI is $y = \dfrac{x}{2} + \dfrac{5}{4}$

The General Solution is

$$y = Ae^{2x} + Be^x + \frac{x}{2} + \frac{5}{4}$$

Example

Find the general solution of the equation

$$\frac{d^2y}{dx^2} - 10\frac{dy}{dx} + 25y = e^{5x}$$

and the particular solution for y=0 and dy/dx = 5 when x=0

$$\frac{d^2y}{dx^2} - 10\frac{dy}{dx} + 25y = e^{5x}$$

has auxillary equation

$m^2 - 10m + 25 = 0$

whose discriminant is

$100 - (4 \times 1 \times (25)) = 0$

∴ *real* and equal roots exist

$m^2 - 10m + 25 = 0$
$\Rightarrow (m-5)(m-5) = 0$
$\Rightarrow m = 5$

The complementary function is
$y = Ae^{5x} + Bxe^{5x}$

For the PI, try $y = Ce^{5x}$
This is already a term in the CF
try $y = Cxe^{5x}$
This is already a term in the CF
try $y = Cx^2e^{5x}$

$$\therefore \frac{dy}{dx} = 2Cxe^{5x} + 5Cx^2e^{5x}$$

$$\frac{d^2y}{dx^2} = 2Ce^{5x} + 10Cxe^{5x} + 10Cxe^{5x} + 25Cx^2e^{5x}$$

the particular solution is $y = 5xe^{5x} + \dfrac{x^2e^{5x}}{2}$

$$\frac{d^2y}{dx^2} - 10\frac{dy}{dx} + 25y - e^{5x}$$

$$2Ce^{5x} + 20Cxe^{5x} + 25Cx^2e^{5x}$$
$$\quad - 10(2Cxe^{5x} + 5Cx^2e^{5x}) + 25(Cx^2e^{5x}) = e^{5x}$$

$\Rightarrow 2Ce^{5x} = e^{5x}$

Equating co-efficients,

$\quad 2C = 1$

$\Rightarrow$ C=1/2

PI is $y = 1/2x^2e^{5x}$

The General Solution is

$$y = Ae^{5x} + Bxe^{5x} + \frac{x^2e^{5x}}{2}$$

NUMBER THEORY

GPS 1.4: Applying algebraic skills to number theory.

THE DIVISION ALGORITHM

For all positive integers a and b, where b ≠ 0,

$$a = qb + r$$

$$\frac{a}{b} = q + \frac{r}{b}$$

a divided by b gives a quotient and remainder.
The quotient q and remainder r are integers

$$0 \leq r < |b|$$

Examples

Use the division algorithm to find the quotient and remainder, when a = 158 and b = 17

$$a = qb + r$$
$$158 = 9 \times 17 + 5$$

Convention uses a dot • to show multiplication instead of a ×
This is fine, since we are purely dealing with integers. No decimals are involved

$$158 = 9 \cdot 17 + 5$$
so $q = 9$ and $r = 5$

THE EUCLIDEAN ALGORITHM

This uses the division algorithm to:-
- find the greatest common divisor (gcd)   [aka highest common factor (hcf)],

> The greatest common denominator (gcd)
> of two integers a and b , at least one of which $\neq 0$
> is found when the remainder in the division algorithm is zero
> $$a = qb + r$$
> so $\quad a = qb + 0$

- find the lowest common multiple (lcm) of two numbers,

> $$lcm = \frac{ab}{(a,b)}$$

- reduce a fraction to its simplest form   (just divide top and bottom by the gcd)

- find relatively prime (coprime) integers . These occur when the gcd (a,b) = 1

- solve equations of the form   gcd (a,b) = ax + by.

If $ax + by = d$ where $\gcd(a,b) = d$
then

$$x_n = x_0 + \left(\frac{b}{d}\right)m$$

$$y_n = y_0 + \left(\frac{a}{d}\right)m$$

describes the general solution x_n, y_n
when the particular solutions x_0, y_0 are known.

Examples

Find the gcd of 135 and 1780

$$a = qb + r$$
$$1780 = 13 \cdot 135 + 25$$
Now continue, replacing *a* with *b* and *b* with *r*
$$135 = 5 \cdot 25 + 10$$
$$25 = 2 \cdot 10 + 5$$
$$10 = 2 \cdot 5 + 0$$
$$\gcd(135, 1780) = 5$$

Find the lcm of 135 and 1780

$$lcm = \frac{ab}{(a,b)}$$
$$= \frac{135 \cdot 1780}{5}$$
$$= \frac{240300}{5}$$
$$= 48060$$
$$\operatorname{lcm}(135, 1780) = 48060$$

Reduce the fraction 1480/128600 to its simplest form

$$a = qb + r$$
$$128600 = 86 \cdot 1480 + 1320$$
$$1480 = 1 \cdot 1320 + 160$$
$$1320 = 8 \cdot 160 + 40$$
$$160 = 4 \cdot 40 + 0$$

$$\gcd(128600, 1480) = 40$$

Divide top and bottom by the gcd

$$\frac{1480 \div 40}{128600 \div 40} = \frac{37}{3215}$$

Show that 34 and 111 are co prime

$$a = qb + r$$
$$111 = 3 \cdot 34 + 9$$
$$34 = 3 \cdot 9 + 7$$
$$9 = 1 \cdot 7 + 2$$
$$7 = 3 \cdot 2 + 1$$
$$2 = 2 \cdot 1 + 0$$

$$\gcd(34, 111) = 1$$

∴ 34 and 111 are coprime.

Solve 34x + 111y = 1, where x and y are integers

$$a = qb + r$$
$$111 = 3 \cdot 34 + 9$$
$$34 = 3 \cdot 9 + 7$$
$$9 = 1 \cdot 7 + 2$$
$$7 = 3 \cdot 2 + 1$$
$$2 = 2 \cdot 1 + 0$$

change subject to *r* and substitute

Start with the second last equation and work backwards

$$\begin{aligned}
1 &= 7 - 3 \cdot 2 \\
&= 7 - 3 \cdot (9 - 1 \cdot 7) \\
&= 7 - 3 \cdot 9 + 3 \cdot 7 = 4 \cdot 7 - 3 \cdot 9 \\
&= 4 \cdot (34 - 3 \cdot 9) - 3 \cdot 9 = 4 \cdot 34 - 15 \cdot 9 \\
&= 4 \cdot 34 - 15 \cdot (111 - 3 \cdot 34) \\
&= 49 \cdot 34 - 15 \cdot 111
\end{aligned}$$

compare to original

$$34x + 111y = 1$$
$$\Rightarrow x = 49 \quad y = -15$$

DIOPHANTINE EQUATIONS

These are of the form

$$ax^n + by^n = c^n$$
where all numbers are integers

$ax + by = c$

has a solution only if the gcd is a factor of c

To solve
1) Find $\gcd(a,b) = d$
then $d \mid c$
so $c = dn$ for some integer n.
Express c in terms of d

2) Express d in the form $d = as + bt$ for some integers s and t

3) Multiply by n to get $x = sn \quad y = tn$

Example

Solve the linear Diophantine Equation
69x +27y = 1332, if it exists.

$$69x + 27y = 1332$$

Find the gcd of 69 and 27
$$69 = 2 \cdot 27 + 15$$
$$27 = 1 \cdot 15 + 12$$
$$15 = 1 \cdot 12 + 3$$
$$12 = 4 \cdot 3 + 0$$
gcd(69, 27) = 3
since 3|1332, a solution exists
$$c = dn \Rightarrow 1332 = 3n \Rightarrow n = 444$$

$$3 = 15 - 1 \cdot 12$$
$$= 15 - 1 \cdot (27 - 1 \cdot 15) = 15 - 1 \cdot 27 + 1 \cdot 15 = 2 \cdot 15 - 1 \cdot 27$$
$$= 2 \cdot (69 - 2 \cdot 27) - 1 \cdot 27$$
$$= 2 \cdot 69 - 4 \cdot 27 - 1 \cdot 27$$
$$= 2 \cdot 69 - 5 \cdot 27$$
$$d = 69s + 27t$$
$$\Rightarrow s = 2 \quad t = -5$$

Find the positive integer values of x and y that satisfy
69x + 27y = 1332.

From above, a solution exists

$$69x + 27y = 1332$$
$$\gcd(69, 27) = 3$$
$$23x + 9y = 444$$

solving for x

$$x = \frac{444 - 9y}{23}$$
$$= 19\frac{7}{23} - \frac{9y}{23}$$
$$= 19 - \frac{9y}{23} + \frac{7}{23}$$
$$= 19 - \frac{9y - 7}{23}$$
$$\therefore \frac{9y - 7}{23} \leq 18$$
$$\Rightarrow y \leq \frac{23 \cdot 18 + 7}{9}$$
$$\Rightarrow y \leq \frac{421}{9} \leq 46\frac{7}{9}$$
$$0 < y \leq 46, \quad \Delta y = 9$$

Lowest possible is
$$y = 11$$
thus $x = \frac{444 - 99}{23} = 15$

Alternatively, solving for y
$$y = \frac{144 - 23x}{9}$$
$$= 49 - \frac{23x - 3}{9}$$
$$\therefore \frac{23x - 3}{9} \leq 48$$
$$\Rightarrow x \leq \frac{48 \cdot 9 + 3}{23}$$
$$\Rightarrow x \leq \frac{435}{23} \leq 18\frac{21}{23}$$
$$0 < x \leq 18, \quad \Delta x = 23$$

Lowest possible answers
$$x = 15$$
$$y = 49 - \frac{23 \cdot 15 - 3}{9}$$
$$= 49 - 38$$
$$y = 11$$

check
$$69x + 17y = 69 \cdot 15 + 17 \cdot 11 = 1035 + 187 = 1332$$

PYTHAGOREAN TRIPLES

$$\boxed{ax^2 + by^2 = c^2}$$

To find these,
Pick an odd positive number
Divide its square into two integers which are as close to being equal as is possible.

e.g. $7^2 = 49 = 24 + 25$ gives triples 7, 24, 25
$7^2 + 24^2 = 25^2$

Alternatively, pick any even integer n
triples are $2n$, $n^2 - 1$ and $n^2 + 1$
e.g. picking 8 gives 16, 63 and 65 and $16^2 + 63^2 = 65^2$

FERMAT'S THEOREM

$$\boxed{ax^n + by^n = c^n, \quad n > 2 \\ \text{cannot be solved with all as integers}}$$

NUMBER BASES

To convert a number into a different base, use the Division Algorithm, taking b as the required base.

In Hexadecimal, the letters A to F are used to represent the numbers from 10 to 15

A =10 B = 11 C = 12 D = 13 E = 14 F = 15

Examples
Convert 36 into hexadecimal

Convert 36 into binary
$$a = qb + r$$
$$36 = 18 \bullet 2 + 0$$

Now continue, replacing a with q
$$18 = 9 \bullet 2 + 0$$
$$9 = 4 \bullet 2 + 1$$
$$4 = 2 \bullet 2 + 0$$
$$2 = 1 \bullet 2 + 0$$
$$1 = 0 \bullet 2 + 1$$

Read the remainder upwards
$$36 = 100100 \quad \text{in base 2}$$

The base is often shown as a subscript
$$36 = 100100_2$$

$$a = qb + r$$
$$36 = 2 \bullet 16 + 4$$
$$2 = 0 \bullet 16 + 2$$

$$36 = 24 \quad \text{in base 16}$$

Convert 503793 into hexadecimal.

MATHEMATICAL PROOF

GPS 1.5 Applying algebraic and geometric skills to methods of proof.

Apps 1.3 Applying algebraic skills to summation and mathematical proof.

PROOF

Mathematics is largely based on logic – a structure of rules used for reasoning.

Most of maths is written as statements : words or symbols which contain information.

A statement may be true or false, and is made from axioms, assumptions and arguments.

An axiom is a statement which is assumed to be true, and is used to then develop a system. All logical systems must state its axioms.

An assumption, or premise, is a statement, which may in reality be true or false but is taken to be true for the argument which follows.

An argument is a set of statements which use logic to show how one particular statement occurs.

A proposition is a statement whose truth is to be shown by the use of an argument.

A conjecture is a proposition which appears to be true, but has not yet been proved.

A theorem is a statement which has been proved to be true.

A lemma is a theorem which is used in the proof of another theorem. A lemming is a small rodent which jumps of cliffs.

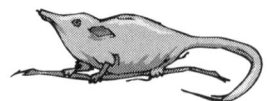

The converse of a theorem or statement takes the conclusion as the starting point, and the starting point as the conclusion.

The inverse of a statement changes its polarity.

The contrapositive of a statement is the negation of the converse. The negation of a statement is made by putting a NOT in it!

Example

If it rains today, we shall have cake tomorrow.

Inverse: If it does not rain today, we shall not have cake tomorrow.

Converse: we shall have cake tomorrow if it rains today.

Contrapositive : we shall not have cake tomorrow if it does not rain today.

A proof is a sequence of statements which lead to the establishment of the truth of one final statement.

A proof may be valid or invalid, depending on whether or not the arguments used are correct. One counter-example will show a statement to be invalid, which will make the proof invalid.

Proof by Exhaustion

Here, a proof is made by showing that the statement holds for every single possible case.

PROOF HAS ITS OWN NOTATION:-

$\therefore$ therefore

$\because$ because or since

$\exists$ there exists

$\ni$ contains as member

$\in$ is an element of

$\notin$ is not an element of

$\forall$ for all

$\neg$ Not

$\wedge$ And

$\vee$ Or

$\Rightarrow$ Implies

$\Leftarrow$ Is implied by

$\Leftrightarrow$ is equivalent to, if and only if

$b|a$ b divides a

so $a = kb$, $k \in W$

DIRECT PROOF

This is proof in which all the assumptions used are true, and all the arguments are valid. A series of linked implications lead from a given statement to a declared goal. The original statement is given to be true, implying the declared goal is also true.

Examples

Given x is an even number, prove x^2 is an even number.

x is even
$\Rightarrow x = 2k, \quad k \in W$
$\Rightarrow x^2 = (2k)^2$
$\Rightarrow x^2 = 4k^2$
$\Rightarrow x^2 = 2(2k^2)$
$\Rightarrow x^2 = 2(2k^2)$

x is even (given)
$\therefore x^2$ is even

Prove that the arithmetic series with first term a and common difference d can be summed as shown:- $S_n = \dfrac{n}{2}(2a + (n-1)d)$

$S_n = \sum_{r=1}^{n} a + (n-1)d$

$= a + (a+d) + (a+2d) + (a+3d) + \ldots + (a+(n-2)d) + (a+(n-1)d)$

but

$S_n = (a+(n-1)d) + (a+(n-2)d) + \ldots + (a+3d) + (a+2d) + (a+d) + a$

Adding

$2S_n = a + (a+(n-1)d) + (a+d) + (a+(n-2)d) + \ldots + (a+(n-2)d) + (a+2d) + (a+(n-1)d) + a$

$= (2a+(n-1)d) + (2a+(n-1)d) + \ldots + (2a+(n-1)d) + (2a+(n-1)d)$

$= n(2a+(n-1)d)$

$\therefore S_n = \dfrac{n(2a+(n-1)d)}{2}$

INDIRECT PROOF (AKA PROOF BY CONTRADICTION)

One assumption is made which is the negation of the statement to be proved. Valid arguments are then used to arrive at a statement which is clearly false.

This contradiction then makes the original assumption false, thus making the statement to be proved true.

Example

Show by contradiction that the square root of any prime number is irrational.

Assume that prime number (p) exists with a rational square root.

Then

$$\sqrt{p} = \frac{a}{b}$$

$$\Rightarrow p = \frac{a^2}{b^2}$$

$\Rightarrow b^2 = 1$, since p is an integer

$\therefore p = a^2$

$\Rightarrow p$ cannot be a prime number

A contradiction exists,
thus the square root of any prime number is irrational.

PROOF BY CONTRAPOSITIVE

Similar to above, however this time the contrapositive is used:- instead of a contradiction, the aim is to prove that if statement A implies statement B, then NOT B implies NOT A.

Example

Prove that if $x=89$ then x is not divisible by 3.

If x is divisible by 3, then it has no remainder
$\Leftrightarrow$ If x has a remainder upon division by 3,
    then x is not divisible by 3.

$89 \div 3$ has a remainder
$\Rightarrow 89 \not{\div} 3$
$\Rightarrow x \not{\div} 3$

PROOF BY INDUCTION

A conjecture is made and shown to be true for numbers n and n+1.
Then it is shown to be true for a specific low number.
This then leads to the conclusion that it will be true for all values of n ≥ the specific value proved in step 2.

Note: The order of these steps does not matter

Examples

Prove by induction that $2^n > n$, $\forall n \in \mathbb{N}$

Assume it is true when $n = k$

Then $2^k > k$
$\Rightarrow 2 \times 2^k > 2k$
$\Rightarrow 2^{k+1} > 2k$
$\Rightarrow 2^{k+1} > k + k$
$\Rightarrow 2^{k+1} > k + 1$
$\Rightarrow$ the statement is valid for $n = k+1$

Let n=1

then 2 > 1 so the statement is valid.

The statement is true for n=k (given), n=k+1 and n=1

∴ by induction, $2^n > n$, $\forall n \geq 1, n \in \mathbb{N}$

Prove by induction that, for all positive integers n,
$1+2+3+...+n = n(n+1)/2$

Let T be the solution set.
Assume it is true when n = 1
Then $1 = 1(1+1)/2$
RHS $1(1+1)/2 = 2/2 = 1$
∴ the statement is valid for n=1, so $1 \in T$

Let $n = k$, then
$1+2+3+...+k = k(k+1)/2$

Let $n = k+1$, then
$1+2+3+...+k+(k+1) = (k+1)(k+2)/2$
LHS
$1+2+3+...+k+(k+1)$
$= (1+2+3+...+k)+(k+1)$
$= k(k+1)/2+(k+1)$
$= \dfrac{k^2+k+2k+2}{2}$
$= \dfrac{(k+1)(k+2)}{2} = $ RHS

so if $k \in T$, $k+1 \in T$
the statement is valid.

∴ by induction,
$1+2+3+...+n = n(n+1)/2 \quad \forall n \in \mathbb{Z}.>0$

Prove by induction that, for all positive integers n,
$1+3+6+10+...+n(n+1)/2 = n(n+1)(n+2)/6$

Let T be the solution set.
Assume it is true when n = 1
Then $1 = 1(1+1)(1+2)/6$
RHS $1(1+1)(1+2)/6 = (2\times 3)/6 = 1$
∴ the statement is valid for n=1, $1 \in T$

Let $n = k$, then
$1+3+6+10+...+k(k+1)/2 = k(k+1)(k+2)/6$

Let $n = k+1$, then
$1+3+6+10+...+k(k+1)/2+(k+1)(k+2)/2 = (k+1)(k+2)(k+3)/6$
LHS
$1+3+6+10+...+k(k+1)/2+(k+1)(k+2)/2$
$= (1+3+6+10+...+k(k+1)/2)+(k+1)(k+2)/2$
$= k(k+1)(k+2)/6+(k+1)(k+2)/2$
$= \dfrac{k(k+1)(k+2)+3(k+1)(k+2)}{6}$
$= \dfrac{(k+1)(k+2)(k+3)}{6} = $ RHS

so if $k \in T$, $k+1 \in T$
the statement is valid.

∴ by induction,
$1+3+6+10+...+n(n+1)/2 = n(n+1)(n+2)/6$

Prove by induction that, for all positive integers n,
$$\sum_{r=1}^{n}(2r-1)=n^2$$

Let T be the solution set.
Assume it is true when n = 1
Then $1 = 1^2$
∴ the statement is valid for n=1, $1 \in T$

Let $n = k$, then
$$\sum_{r=1}^{k}(2r-1)=k^2$$

Let $n = k+1$, then
$$\sum_{r=1}^{k+1}(2r-1)=\sum_{r=1}^{k}(2r-1)+(2(k+1)-1)=(k+1)^2$$

LHS
$$\sum_{r=1}^{k}(2r-1)+(2(k+1)-1)$$
$= k^2 + (2(k+1)-1)$
$= k^2 + 2k + 1$
$= (k+1)(k+1)$
$= (k+1)^2 =$ RHS

so if $k \in T$, $k+1 \in T$
the statement is valid.

∴ by induction,
$$\sum_{r=1}^{n}(2r-1)=n^2$$

Prove by induction that, for all positive integers n,
$(\cos\theta + i\sin\theta)^n = \cos n\theta + i\sin n\theta$

Let T be the solution set.
Assume it is true when n = 1
Then $\cos\theta + i\sin\theta = \cos\theta + i\sin\theta$
∴ the statement is valid for n=1, $1 \in T$

Let $n = k$, then
$(\cos\theta + i\sin\theta)^k = \cos k\theta + i\sin k\theta$

Let $n = k+1$, then
$(\cos\theta + i\sin\theta)^{k+1} = \cos(k+1)\theta + i\sin(k+1)\theta$

LHS
$(\cos\theta + i\sin\theta)^{k+1}$
$= (\cos\theta + i\sin\theta)^k (\cos\theta + i\sin\theta)$
$= (\cos k\theta + i\sin k\theta)(\cos\theta + i\sin\theta)$
$= \cos k\theta \cos\theta + i\sin k\theta \cos\theta + i\sin\theta \cos k\theta - \sin\theta \sin k\theta$
$= \cos(k+1)\theta + i\sin(k+1)\theta = $ RHS

so if $k \in T$, $k+1 \in T$
the statement is valid.

∴ by induction,
$(\cos\theta + i\sin\theta)^n = \cos n\theta + i\sin n\theta$

Prove by induction that the integer
$(2n-1)^2 -1$ is divisible by 8, $\forall n \in N$

Let T be the solution set.
Assume it is true when $n = 1$
then $(2n-1)^2 -1 = 0/8 = 0$
so the statement is valid.

Assume it is true when $n = k$
Then $8 | (2k-1)^2 -1$
$\Rightarrow (2k-1)^2 -1 = 8b$ for some integer b

Let $n = k+1$, then
$(2(k+1)-1)^2 -1 = 8c$ for some integer c
LHS
$(2(k+1)-1)^2 -1 = (2k+1)^2 - 1$
$= 4k^2 + 4k = 4k^2 - 4k + 8k$
$= (2k-1)^2 - 1 + 8k = 8b + 8k$
$= 8(b+k) = 8c$ where $c = a + k$
=RHS

$c = a + k$ is an integer
so if $k \in T$, $k+1 \in T$
the statement is valid.

$\therefore$ by induction,
$(2n-1)^2 -1$ is divisible by 8, $\forall n \in N$

Prove by induction that, for all positive integers n,

$$\sum_{r=1}^{n} \frac{1}{(3r-2)(3r+1)} = \frac{n}{3n+1}$$

Let T be the solution set.
Assume it is true when n = 1

Then $\frac{1}{(3-2)(3+1)} = \frac{1}{4}$

RHS $\frac{1}{(3-2)(3+1)} = \frac{1}{1 \times 4} = \frac{1}{4}$

∴ the statement is valid for n=1, $1 \in T$

Let $n = k$, then

$$\sum_{r=1}^{k} \frac{1}{(3r-2)(3r+1)} = \frac{k}{3k+1}$$

Let $n = k+1$, then

$$\sum_{r=1}^{k+1} \frac{1}{(3r-2)(3r+1)} = \frac{k+1}{3(k+1)+1} = \frac{k+1}{3k+4}$$

LHS

$$\sum_{r=1}^{k+1} \frac{1}{(3r-2)(3r+1)} = \sum_{r=1}^{k} \frac{1}{(3r-2)(3r+1)} + \frac{1}{(3(k+1)-2)(3(k+1)+1)}$$

$$= \sum_{r=1}^{k} \frac{1}{(3r-2)(3r+1)} + \frac{1}{(3k+1)(3k+4)}$$

$$= \frac{k}{3k+1} + \frac{1}{(3k+1)(3k+4)}$$

$$= \frac{k(3k+4)+1}{(3k+1)(3k+4)}$$

$$= \frac{3k^2 + 4k + 1}{(3k+1)(3k+4)}$$

$$= \frac{(3k+1)(k+1)}{(3k+1)(3k+4)}$$

$$= \frac{(k+1)}{(3k+4)} = \text{RHS}$$

so if $k \in T$, $k+1 \in T$
the statement is valid.

∴ by induction,

$$\sum_{r=1}^{n} \frac{1}{(3r-2)(3r+1)} = \frac{n}{3n+1}$$

USEFUL FORMULAE

SET NOTATION

Notation	Examples
$\in$ means " is an element of "	$2 \in \{1,2,3,4\}$
$\notin$ means " is not an element of "	$2 \notin \{1,3,5\}$
$\cup$ means "union of set"	$\{1,2,3\} \cup \{3,4,8\} = \{1,2,3,4,8\}$
$\cap$ means "intersection of set"	$\{1,2,3\} \cap \{3,4,8\} = \{3\}$
$\subset$ means "subset of set"	$\{1,2,3\} \subset \{1,2,3,4,8\}$
$\supset$ means "superset of set"	$\{1,2,3,4,8\} \supset \{1,2,3\}$
$\not\subset$ means "not a subset of set"	$\{10,20,30\} \not\subset \{1,2,4,8\}$
$\emptyset$ or $\{\}$ denotes the empty set	

Common Sets

$\mathbb{N}$ is the set of natural numbers used in counting: $\{1,2,3,4.....\infty\}$

$\mathbb{W}$ is the set of whole numbers: $\{0,1,2,3,4.....\infty\}$

$\mathbb{Z}$ is the set of integers:- positive and negative whole numbers $\{-\infty...-4,-3,-2,-,1,0,1,2,3,4......\infty\}$

$\mathbb{W} \subset \mathbb{Z}$

$\mathbb{R}$ is the set of real numbers $\{-\infty,........\infty\}$
This includes all numbers, rational and irrational.
So $\mathbb{Q} \subset \mathbb{R}$
Positive numbers are real numbers which are greater than zero.
Negative numbers are real numbers which are less than zero.
$\mathbb{N} \subset \mathbb{W} \subset \mathbb{Z} \subset \mathbb{Q} \subset \mathbb{R}$

$\mathbb{Q}$ is the set of rational numbers or quotients.
These are all numbers which can be expressed as a fraction, $\frac{a}{b}$ where both a and b are integers, and b is not zero.
Since integers can be expressed as a fraction with denominator = 1,
$\mathbb{Z} \subset \mathbb{Q}$

$\mathbb{C}$ is the set of complex numbers, $a+bi$
where a and b are real and i is the imaginary number $\sqrt{-1}$

DERIVATIVES

f(x)=	f'(x)=
x^n	nx^{n-1}
ax^n	anx^{n-1}
$(x+b)^n$	$n(x+b)^{n-1}$
$(ax+b)^n$	$an(ax+b)^{n-1}$
Exponentials and logarithmic functions	
e^x	e^x where $e = \lim_{n \to \infty}\left(1+\frac{1}{n}\right)^n$
e^u where $u = f(x)$ (chain rule)	$\frac{d}{dx}e^u = e^u \frac{du}{dx}$
b^x	$\dfrac{b^x}{\log_b e}$
$\log_b x$	$\dfrac{1}{x}\log_b e$
$\log_e x = \ln x$	$\dfrac{1}{x}$

Trig Functions	Don't forget to apply chain rule
sinx	cosx
cosx	−sinx
sinax	acosax
sinu where u = f(x) (chain rule)	$\dfrac{d}{dx}\sin u = \cos u \dfrac{du}{dx}$
cosu where u = f(x) (chain rule)	$\dfrac{d}{dx}\cos u = -\sin u \dfrac{du}{dx}$
cosax	−asinax
tanx	sec²x
tanu where u = f(x) (chain rule)	$\dfrac{d}{dx}\tan u = \sec^2 u \dfrac{du}{dx}$
cosecx	−cosecxcotx
secx	secxtanx
cotx	−cosec²x
Inverse Trig Functions	
arcsinx (sin⁻¹)	$\dfrac{1}{\sqrt{1-x^2}}$
arccosx (cos⁻¹)	$\dfrac{-1}{\sqrt{1-x^2}}$
arctanx (tan⁻¹)	$\dfrac{1}{1+x^2}$
Sum Rule f(x) = g(x) + h(x)	f'(x) = g'(x) + h'(x) (f + g)' = f' + g'

Constant Multiple Rule	$(Kf)' = Kf'$ K constant
	$\dfrac{d}{dx}(cu) = c\dfrac{du}{dx}$ c is a constant
Chain Rule $h(x) = g(f(x))$	$h'(x) = g'(f(x)) \times f'(x)$ or in Leibnitz notation $\dfrac{dy}{dx} = \dfrac{dy}{du} \times \dfrac{du}{dx}$ $(f \circ g)' = (f' \circ g) \times g'$
Product Rule $f(x) = g(x).h(x)$	$f'(x) = g'(x).h(x) + g(x).h'(x)$ $\dfrac{d}{dx}(uv) = u\dfrac{dv}{dx} + v\dfrac{du}{dx}$ $\dfrac{d}{dx}(uvw) = uv\dfrac{dw}{dx} + uw\dfrac{dv}{dx} + vw\dfrac{du}{dx}$
Quotient rule $f(x) = \dfrac{g(x)}{h(x)}$	$f'(x) = \dfrac{g'(x)h(x) - g(x)h'(x)}{(h(x))^2}$ $\dfrac{d}{dx}\left(\dfrac{u}{v}\right) = \dfrac{v\dfrac{du}{dx} - u\dfrac{dv}{dx}}{v^2}$

STANDARD INTEGRALS

$\int x^n = \dfrac{x^{n+1}}{n+1} + C$	
$\int ax^n = \dfrac{ax^{n+1}}{n+1} + C$	$n \neq -1$
Sum Rule $\int (f(x) + g(x))dx = \int f(x)dx + \int g(x)dx$	
Constant Multiplier Rule $\int kf(x)dx = k\int f(x)dx, \quad k \text{ is a constant}$	
$\int (ax+b)^n dx = \dfrac{(ax+b)^{n+1}}{a(n+1)} + c, \quad n \neq -1$	
Integration by Parts $\int u\,dv = uv - \int v\,du$	
Exponentials and logarithms	
$\int \dfrac{1}{x}dx = \ln|x| + C$	
$\int \dfrac{dx}{a+bx} = \dfrac{1}{b}\ln|a+bx| + C$	
$\int e^x dx = e^x + c$	

Trig Functions
$\int \sin x\, dx = -\cos x + C$
$\int \cos x\, dx = \sin x + C$
$\int \sin(ax+b)\, dx = -\dfrac{1}{a}\cos(ax+b) + c$
$\int \cos(ax+b)\, dx = \dfrac{1}{a}\sin(ax+b) + c$
$\int \tan x\, dx = \int \dfrac{\sin x}{\cos x}\, dx = -\ln|\cos x| + c$
$\int \cot x\, dx = \ln|\sin x| + c$
$\int \sec^2 x\, dx = \tan x + C$
$\int \tan^2 x\, dx = \int (\sec^2 x - 1)\, dx = \tan x - x + c$

Inverse Trig Functions
$\int \dfrac{1}{\sqrt{1-x^2}}\, dx = \sin^{-1} x + C$ (arc sin)
$\int \dfrac{1}{\sqrt{a^2 - x^2}}\, dx = \sin^{-1}\left(\dfrac{x}{a}\right) + C$
$\int \dfrac{1}{1+x^2}\, dx = \tan^{-1} x + C$ (arc tan)
$\int \dfrac{1}{a^2 + x^2}\, dx = \dfrac{1}{a}\tan^{-1}\left(\dfrac{x}{a}\right) + C$
$\int \sqrt{1-x^2}\, dx = \dfrac{1}{2}\left(\sin^{-1} x + x\sqrt{1-x^2}\right) + C$

Common Forms

$$\int f(ax+b)dx = \frac{1}{a}F(ax+b)+C$$

$$\int \frac{f'(x)}{f(x)}dx = \ln|f(x)|+C$$

$$\int f'(x)f(x)dx = \frac{1}{2}(f(x))^2 + C$$

Rectilinear motion (straight line motion)

velocity = rate of change of displacement with time

$$v = \frac{ds}{dt}$$

acceleration = rate of change of velocity with time

$$a = \frac{dv}{dt} = \frac{d^2s}{dt^2}$$

$$a = \frac{dv}{dt} \quad , \quad v = \int a\,dt$$

$$v = \frac{ds}{dt} \quad , \quad s = \int v\,dt$$

Volumes of Solid of Revolution

$$V = \int_c^d \pi x^2 dy \qquad\qquad V = \int_a^b \pi y^2 dx$$

TRIG FORMULAE

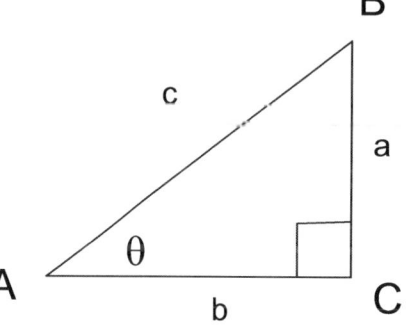

Sine of A

$$SinA = \frac{Opposite}{Hypotenuse}$$

$$= \frac{a}{c}$$

Cosine of A

$$CosA = \frac{Adjacent}{Hypotenuse}$$

$$= \frac{b}{c}$$

****

Cosecant of A $= \dfrac{1}{SinA}$

$$Co\sec A = \frac{Hypotenuse}{Opposite}$$

$$Co\sec A = \frac{c}{a}$$

Secant of A $= \dfrac{1}{CosA}$

$$SecA = \frac{Hypotenuse}{Adjacent}$$

$$SecA = \frac{c}{b}$$

****

Tangent of A

$$TanA = \frac{Opposite}{Adjacent}$$

$$= \frac{a}{b}$$

Cotangent of A $= \dfrac{1}{TanA}$

$$CotA = \frac{Adjacent}{Opposite}$$

$$= \frac{b}{a}$$

****

$$\operatorname{Cosec} A = \frac{1}{SinA} \qquad SecA = \frac{1}{CosA} \qquad CotA = \frac{1}{TanA}$$

$$\operatorname{Cosec}^2 A = \frac{1}{Sin^2 A} \qquad Sec^2 A = \frac{1}{Cos^2 A} \qquad Cot^2 A = \frac{1}{Tan^2 A}$$

$$\cos^2\theta + \sin^2\theta = 1$$
$$\sec^2\theta - \tan^2\theta = 1$$
$$\operatorname{cosec}^2\theta - \cot^2\theta = 1$$

$$\sin(-\theta) = -\sin\theta \qquad \operatorname{cosec}(-\theta) = -\operatorname{cosec}\theta$$
$$\cos(-\theta) = \cos\theta \qquad \sec(-\theta) = \sec\theta$$
$$\tan(-\theta) = -\tan\theta \qquad \cot(-\theta) = -\cot\theta$$

COMPOUND FORMULAE

$\sin(A+B) = \sin A \cos B + \cos A \sin B$

$\sin(A-B) = \sin A \cos B - \cos A \sin B$

$\cos(A+B) = \cos A \cos B - \sin A \sin B$

$\cos(A-B) = \cos A \cos B + \sin A \sin B$

$\tan(A+B) = \dfrac{\tan A + \tan B}{1 - \tan A \tan B} \qquad A+B \neq \dfrac{(2n+1)\pi}{2}$

$\tan(A-B) = \dfrac{\tan A - \tan B}{1 + \tan A \tan B} \qquad A+B \neq \dfrac{(2n+1)\pi}{2}$

$\cot(A+B) = \dfrac{\cot A \cot B - 1}{\cot B + \cot A}$

$\cot(A-B) = \dfrac{\cot A \cot B + 1}{\cot B - \cot A}$

DOUBLE ANGLE FORMULAE   ALF ANGLE FORMULAE

$\sin 2A = 2\sin A \cos A$

$\sin\dfrac{A}{2} = \pm\sqrt{\dfrac{1-\cos A}{2}}$ + if A/2 is in 1st or 2nd quadrant
− if A/2 is in 3rd or 4th quadrant

$\cos 2A = \cos^2 A - \sin^2 A$
$= 1 - 2\sin^2 A$
$= 2\cos^2 A - 1$

$\cos\dfrac{A}{2} = \pm\sqrt{\dfrac{1+\cos A}{2}}$ + if A/2 is in 1st or 4th quadrant
− if A/2 is in 2nd or 3rd quadrant

$\tan 2A = \dfrac{2\tan A}{1-\tan^2 A}$

$\tan\dfrac{A}{2} = \pm\sqrt{\dfrac{1-\cos A}{1+\cos A}}$ + if A/2 is in 1st or 3rd quadrant
− if A/2 is in 2nd or 4th quadrant

ADDITION FORMULAE   PRODUCTS

$\sin A + \sin B = 2\sin\left(\dfrac{A+B}{2}\right)\cos\left(\dfrac{A-B}{2}\right)$

$\sin A \sin B = \dfrac{\cos(A-B) - \cos(A+B)}{2}$

$\sin A - \sin B = 2\cos\left(\dfrac{A+B}{2}\right)\sin\left(\dfrac{A-B}{2}\right)$

$\sin A \cos B = \dfrac{\sin(A+B) + \sin(A-B)}{2}$

$\cos A + \cos B = 2\cos\left(\dfrac{A+B}{2}\right)\cos\left(\dfrac{A-B}{2}\right)$

$\cos A \cos B = \dfrac{\cos(A-B) + \cos(A+B)}{2}$

$\cos A - \cos B = -2\sin\left(\dfrac{A+B}{2}\right)\sin\left(\dfrac{A-B}{2}\right)$

$\cos A \sin B = \dfrac{\sin(A+B) - \sin(A-B)}{2}$

MULTIPLE ANGLE FORMULAE

$$\sin 3A = 3\sin A - 4\sin^3 A$$

$$\cos 3A = 4\cos^3 A - 3\cos A$$

$$\tan 3A = \frac{3\tan A - \tan^3 A}{1 - 3\tan^2 A}$$

****

$$\sin 4A = 4\sin A \cos A - 8\sin^3 A \cos A$$

$$\cos 4A = 8\cos^4 A - 8\cos^2 A + 1$$

$$\tan 4A = \frac{4\tan A - 4\tan^3 A}{1 - 6\tan^2 A + \tan^4 A}$$

****

$$\sin 5A = 5\sin A - 20\sin^3 A + 16\sin^5 A$$

$$\cos 5A = 16\cos^5 A - 20\cos^3 A + 5\cos A$$

$$\tan 5A = \frac{\tan^5 A - 10\tan^3 A + 5\tan A}{1 - 10\tan^2 A + 5\tan^4 A}$$

BOOKS BY THE AUTHOR

Printed Books

Pocket Revision: Basic Trigonometry

ISBN 978 - 0 - 9576916-2-9

Sudoku Puzzles
ISBN-13: 978-1797866956

Word Sudoku Puzzles
ISBN-13: 978-1796912784

Higher Mathematics Revision Questions: With Worked Solutions
ISBN-13: 978-1799179559

Kindle Books

A variety of books in Kindle format are available.

INDEX

A line of intersection, 162
A plane of intersection, 163
Absolute convergence, 60
Adjoint, 116
amplitude, 31
Applications of Differential Equations, 290
Applications of differentiation, 214
Applications of integration, 297
Approximating roots of an equation Newton's Method, 213
Area between curve and the y-axis, 269
Area between two graphs, 272
Areas enclosed by the graph and the x – axis., 271
Areas under curves, 261
arg z, 31
Argand diagrams, 29
argument, 324
argument of z, 31
Arithmetic series, 51
assumption, 323
Asymptotes, 90
augmented matrix, 121
auxiliary equation, 302
axiom, 323
between a line and a plane, 150
Binomial expansions, 19
Centre of convergence, 62
Closed Intervals, 202
Cofactors, 113
combination, 15
common difference, 41
common ratio, 47
Common Sets, 338
Complex Conjugate, 26
complex plane, 29
Complex roots, 37
conjecture, 324
Constraint equations, 234
contrapositive, 324, 325
converse, 324, 325
Convert into binary, 321
Convert into hexadecimal, 321
Coplanar vectors, 143
coprime, 312
counter-example, 325

Critical Points, 93
cross product, 129
D'Alembert's ratio test, 58
De moivre's, 34
Definite integrals, 267
degree of a differential equation, 284
dependent variable, 227
Derivatives, 339
Determinant of a 3x3 matrix, 111
Determinants, 110
Differentiating Explicit and Implicit Functions, 227
Differentiating Inverse functions, 219
Differentiating Inverse Trig functions, 223
Differentiating parametric equations, 236
Differentiation, 170
Differentiation by First Principles, 178
Differentiation Calculations Refresher, 204
Diophantine Equations, 317
Direct Proof, 327
Direction Ratios and Cosines, 125
distance between parallel planes, 142
distance from a point to a plane, 157
distance from a point to a line, 159
Distinct linear factors, 166
Elementary Row Operations, 121
Elements, 99
Enlargement with scale factor, 109
equations of a line, 145
explicit function, 227
Factorials, 13
Fermat' s Theorem, 320
Fibonacci series, 62
Finding asymptotes, 91
Finding Gradients of Curves, 171
Finding greatest / least value, 214
Finding other derivatives by first principles, 181
Finding Stationary Points, 199
Finding the equation of a tangent, 196
First order Linear Differential Equations, 299
First order process, 81
Fixed Points, 40
Formula for Area between two graphs, 273
Fundamental theorem of calculus, 270
Gaussian Elimination, 121
General Differentiation, 179

General rules, 204
Geometric sequences, 47
Geometric series, 52
Gradients of tangents to curves, 196
greatest common divisor, 312
Higher Derivatives, 203
Implicit functions, 227
Increasing / Decreasing functions, 197
independent variable, 227
Indirect Proof, 328
Infinite Endpoint, 257
Infinite Integrals, 255
Infinite point within the integrand, 259
infinite series, 51
Integrating Factor, 299
Integrating inverse trig functions, 276
Integrating Logarithms and Exponentials, 246
Integrating Trig functions, 245
Integration, 243
Integration and the Area Function, 266
Integration by Parts, 277
Integration by substitution, 247
Integration Common Forms, 250
intersection of three planes, 161
intersection of two lines, 152
intersection of two planes, 155
interval of convergence, 62
invalid, 325
invariant, 109
inverse, 324, 325
Inverse of a square matrix, 117
irreducible quadratic factor, 168
irst-order, 39
Iteration, 74
lemma, 324
Linear Differential Equations, 298
linear recurrence relation, 39
Loci, 32
Logarithmic Differentiation, 230
Logarithms and Exponentials, 212
lower bound, 262
lowest common multiple, 312
MACLAURIN'S SERIES, 65
Matrix Addition& Subtraction, 101
Matrix Equality, 100
Matrix Multiplication, 102

Matrix notation, 100
Matrix Transpose, 105
modulus, 30
Modulus, 86
Nature Tables, 200
Newton Raphson Iteration, 74
Newton's Law of Cooling, 291
notation, 326
Number Bases, 321
Odd and Even functions, 87
order of a differential, 284
orthogonal, 109
Other properties vector product, 128
parametric Equations, 232
Partial Fractions, 165
partial sum, 51
particular integral, 306
Pascal's Triangle, 17
permutation, 15
Permutations and Combinations, 15
Planes in space, 135, 140
Polar form, 33
Power Series, 58
premise, 323
Probability and the binomial theorem, 22
Product of a square matrix and its inverse, 118
Product Rule, 188, 208
proof, 325
Proof by contradiction, 328
Proof by Contrapositive, 329
Proof by Exhaustion, 325
Proof by Induction, 330
proposition, 324
Pythagorean Triples, 320
Quotient Rule, 189, 209
radius of convergence, 62
Rectilinear motion (straight line motion), 344
Reflection and Rotation, 108
relatively prime, 312
repeated linear factor, 167
Right Hand Screw rule., 127
Rocket Science, 290
Roots of a complex number, 35
Rule of false position, 83
Scalar Matrix Multiplication, 102
Scalar Triple Product, 133

Second Derivative, 201
Second order Linear Differential Equations, 302
Second order non – homogeneous Differential Equations, 306
Second order process, 82
series, 39
Sets reminder, 25
Sigma notation, 54
Solving by direct integration, 285
Special matrices, 106
stable fixed point, 40
Standard Integrals, 342
Standard series, 72
statement, 323
statements, 323
stationary point, 198
substitute dx, 248
Sum Rule, 205
sum to infinity, 51
Sum to n terms of a geometric sequence, 49
Sum to n terms of an arithmetic sequence, 43
Taylor's series, 65
tegrating rational functions, 279
The angle between two planes, 141

The binomial expansion and e, 23
The Chain rule, 207
The Chain Rule, 187
The Division Algorithm, 311
The Euclidean Algorithm, 312
theorem, 324
Three lines of intersection, 163
Transformation Matrices, 107
Trig formulae, 345
Trig functions, 210
trigonometric derivatives by first principles, 190
Two lines of intersection, 163
unstable fixed point, 40
upper bound, 262
upper triangular form., 121
Using Taylor's series, 81
valid, 325
variables separable, 287
vector equation of the plane, 143
Vector Product, 127
volume of a parallelepiped, 133
Volumes of Solid of Revolution, 294, 344

Printed in Great Britain
by Amazon